Starting and Developing a Surveying Business: for Sole Traders

Austen Imber

LONDON AND NEW YORK

First published 2005 by Estates Gazette

Published 2014 by Routledge
2 Park Square, Milton Park, Abingdon, Oxon OX14 4RN
711 Third Avenue, New York, NY 10017, USA

Routledge is an imprint of the Taylor & Francis Group, an informa business

Typeset in Palatino by Amy Boyle, Rochester, Kent

ISBN - 978 0 7282 0454 6 (pbk)

Contents

Foreword

Setting up in practice alone is a big step for any chartered surveyor to take, and many factors determine the success, or otherwise, that the venture will enjoy. Strongly performing property markets and the modern ways in which sole traders can operate, make the option of self-employment appear attractive. The potential lifestyle and financial gains do, however, belie the immense challenges ahead for most aspiring sole traders. Uncertainty regarding future work, rejections from potential clients, cash flow constraints and isolation from a team environment are just some of the factors contributing to the low points. For successful self-employed surveyors though, these are far outweighed by the benefits. These include the overall sense of fulfilment that self-employment brings, and smaller matters such as the absence of office politics, and not having to answer to others.

I am particularly pleased to be asked to contribute this foreword to *Starting and Developing a Surveying Business* because of my background. In 1996 I started up in circumstances that many others would not: a 25 year old female, recently married and planning a family, and with surveying experience only in the public sector. My circumstances contrasted with the many considerably more experienced surveyors working in salaried positions in private practice who never take the step, despite quietly contemplating the possibility.

In looking back at the early stages of developing my own business, *Starting and Developing a Surveying Business* would have served me well. Today, it has enabled me to review the opportunities, as well as the threats, faced by the business, and also plan further business development activity. Taxation and personal financial planning are also under review, prompted by the points made in several chapters.

Readers are in good hands with the author, Austen Imber. In *Starting and Developing a Surveying Business*, he combines property knowledge as a chartered general practice surveyor with experience in running his own business, and in advising others. This includes an entrepreneurial eye for the right lines of business to exploit, the effective formulation of strategy, and incisive working methods — all of which help reap financial success.

Mrs Kim Lewis MRICS, The Cass Walker Partnership

Preface

Many surveyors think about the possibility of becoming self-employed, and research often shows being your own boss as the favoured career and lifestyle option.

Common questions asked by prospective sole traders include "how much does it cost to set up", and "how much money will I make?" Start-up costs generally range from less than £1,000 to over £50,000, and profitability ranges from financial losses and the ending of the venture to earnings of £100,000 or more per year. This reflects the multitude of factors which determine fee levels, costs, profit margins and profitability — including the surveyor's age, experience, areas of practice, geographical location, personal qualities, working methods, lifestyle preferences, and the different ways that small businesses operate.

Ways of operating include home-based ventures, run by the surveyor without secretarial support, providing both lifestyle attractions and commercial advantages. Because of advancements in information technology, the traditional surveying business with a shop front or offices, together with receptionists/ secretaries, is only one of many ways in which a surveying business is established. At the extreme, businesses are run from laptops, mobile phones and holiday retreats (in addition to a home base), with the services offered by the business being designed from the outset to achieve this.

Start-up costs and finance need not be a barrier to surveyors wishing to go it alone. Even so, it is hard to make the break from regular employment, and while some surveyors establish their own practice in their 20s and 30s, others make the move only as part of semi-retirement. Critically, surveyors starting up alone need to be able to win business, with business acumen and personal qualities often being more important than technical skills. This need to win business is an important factor to bear in mind when considering the potential profitability of self-employment, and how enticing the lifestyle appears.

The overall emphasis of *Starting and Developing a Surveying Business* is on the drivers of success and profitability of a new sole trader surveying business which are taken account of in its inception. This reflects the importance of the right blend of services, clients, and working methods having particular regard to aspects such as costs, profit margins and chargeable time — rather than only fee levels. Indeed, when reviewing the progress of new surveying businesses, or considering the growth potential of established businesses, a general contentment with simply being busy is often found — with attention not being given to factors such as the scope for rejuvenated business development initiatives to secure higher profile clients and greater fee levels/profit margins, and which services and clients are the most profitable.

The example in Chapter 1 of the surveyor setting up business in his early 30s is a good illustration of the contrast between traditional and home based ventures, the focus on drivers within a business, and the interrelationship between fees, costs, and margins. The surveyor set up to work alone from home in the late 1990s, and reached £100,000 profit/earnings within his first full financial/tax year (winning all his business from scratch). This was in a market which included established small practices/partnerships of considerably more experienced surveyors each achieving £30,000 to £40,000 earnings per year. As mentioned in the section on successful business people in Chapter 1, key qualities include an entrepreneurial eye, and seeing things that others cannot. Strategy and focus are also key themes.

Despite the many different circumstances of surveyors, areas of practice, approaches that a new surveying business could take, and therefore the impossibility of prescribing standard models for success, the various examples and illustrations should help surveyors to dwell on their aspirations and formulate their own individual plans — undertaking the necessary market research, as appropriate. *Starting and Developing a Surveying Business* aims to inspire opportunities for individual surveyors, and in establishing the right mindset for success, avoids being overburdened with matters such as acquiring premises, dealing with suppliers or buying a car for the first time. Sources of further information are provided, such as RICS comprehensive free of charge guide to taking on business premises, and the range of literature produced by the high street banks in respect of banking and financial requirements.

The first four chapters concentrate on market opportunities and how the right concept is formulated. Chapter 1 provides an introduction to self-employment, and enables surveyors to consider the surveying services to be provided. Chapter 2 considers the personal qualities needed, the emotional and motivational aspects of self-employment, and the working methods adopted by sole traders. A summary of financial issues is provided in Chapter 3, and Chapter 4 examines the issues typically covered in a business plan, as well as providing further guidance not considered in previous chapters.

Chapter 5 looks at business status, and the issues to consider as sole trader, partnership or limited company — including ways to enhance the credibility of the business. Also provided is a summary of tax issues affecting small surveying businesses.

Ways of winning instructions are examined in Chapter 6, concentrating on the early development of the business. This enables surveyors to consider the methods which are suitable for their own business, and again reflects the range of ways in which new surveying businesses are established — some incurring no marketing expenditure, while others spend £10,000 or more developing a strong brand and high profile. Chapter 7 provides a broad overview of the issues to consider as the business develops, including options in respect of expansion, or whether the preferred approach (and indeed the most profitable format for the business) is to remain as a sole trader. Chapter 8 builds on the examples included in previous chapters of the many different ways sole traders operate, and provides an illustration of a business which is keen to expand.

The next three chapters consider three key areas of compliance. RICS requirements are summarised in Chapter 9, including those in respect of professional indemnity insurance, clients' accounts, complaints procedures, and a range of other issues regarded as professional ethics. VAT is explained in Chapter 10, and accounting is examined in Chapter 11. Chapters 5, 10 and 11 show how profitability, and the cash flow, of a surveying business is influenced by getting the best out of the tax system — with surveyors' own familiarity with accounting and taxation issues enabling tax efficiencies and financial benefits to be gained.

Brief consideration is given in Chapter 12 to the establishing of property investment and development interests alongside surveying work. Surveyors are ideally placed to enhance earnings in this way, as well as effectively providing a pension through property holdings. The chapter concentrates on issues not commonly found in property text books and journals.

Starting and Developing a Surveying Business does not examine general business skills, such as an understanding of the economy and the financial markets, negotiation skills, presentation techniques, and leadership skills. Similarly, detail is not provided on the ongoing management of a surveying business, strategic review of the business, and succession planning. While the need to consider issues such as pensions, life assurance, mortgages and other aspects of personal and financial planning are highlighted, detailed guidance is not provided. This reflects the importance of surveyors seeking advice specifically in relation to their individual circumstances, and because such areas are not specific to self-employment. It is assumed that surveyors are familiar with the requirements within their many different areas of practice, such as a copy of the Red Book, IT requirements and copyright licences for reproducing plans.

While *Starting and Developing a Surveying Business* aims to provide practical guidance, it is unable to venture into potentially contentious areas such as the effect of age and gender on business dealings, the deployment of particularly aggressive business development techniques, and opportunities which push, if not break, ethical boundaries.

Within a brief summary of RICS and other contacts in Chapter 13, information regarding sector-specific support to surveyors looking to set up their own business is provided, including a concessionary scheme established in view of the publication of *Starting and Developing a Surveying Business*. Details are also provided of free of charge CPD type updates for sole traders and other small surveying businesses.

Thanks are due to *Estates Gazette* editor, Peter Bill and to Estates Gazette Books Commissioning Editor, Alison Richards, plus all involved in the process — including Rebecca Chakraborty of Estates Gazette Books, and Adam Tinworth and

Phil Brown of Estates Gazette. Essential support in proof reading has been provided by Howard Imber, and in production by Audrey Andersson and Amy Boyle. Thanks are also due to the Royal Institution of Chartered Surveyors for confirming the accuracy of Chapter 9, and to Marler Waterhouse for the use of case examples.

Austen Imber
April 2005

Assessing the Opportunities

Self-employment, and starting and developing a surveying business as a sole trader, encompasses many situations, ranging from a new business which grows quickly in profitability and staff numbers, to surveyors undertaking part-time work alongside leisure pursuits — or perhaps even other business interests.

In surveying, the financial rewards are potentially high for those surveyors who exploit gaps in the market, and combine technical proficiency, business instinct and hard work. Others struggle to achieve adequate profitability, and return to salaried employment. A range of guidance on starting a business is available from government departments, banks, business books and specialist magazines, all of which make self-employment appear enticing. However, for a business to succeed, the underlying long term business case has to be sound.

It is not necessarily the most gifted or experienced surveyor who is best suited to self-employment, nor the one most likely to succeed. One surveyor does better entering an already crowded market because of the business development initiatives deployed. Another offering niche services with great potential may struggle to win business or manage the venture effectively.

As mentioned in the preface, there are many ways in which sole traders and small surveying businesses operate, and surveyors have a range of circumstances, fields of practice, expertise and personal aspirations which affect the fee levels achievable, cost base, profitability, etc. Although it is not possible in *Starting and Developing a Surveying Business* to provide prescriptive advice which fits every individual surveyor's circumstances, the comments in the early chapters relate to the majority of surveyors considering setting up their own business, and help in the formulation of individual plans.

This chapter examines the issues which prompt surveyors' thoughts of going it alone, and provides insights into the highs and lows of self-employment. It considers the scope to work from home, together with the necessity of secretarial support, and worked examples show why profit margins and profit, not just fee levels, are key factors to consider. An explanation is provided of the various issues which determine the line of business that the sole trader will pursue — including skills review, market analysis, specialist services, opportunities in crowded markets, level of market entry, how business will be won, rolling out services, cash flow and the defensiveness of fee lines, securing a mix of clients and the complexity and diversity of work taken on. Comment is then made on issues relating to existing employment, partnerships, and advisers/suppliers. Key points are provided on what makes business people successful, and the chapter ends with examples of new surveying businesses.

Initial considerations

Surveyors' thoughts of running their own business are easily prompted by factors other than its commercial merits. Examples include discontentment with a current employer, the attitude of a manager, the journey to work, low pay and long hours. The attractions of self-employment include working from home, flexible hours, visiting the gym in the day, and stopping work early in the evening to enjoy social activities. There is sole control over the direction of the business, and approvals from managers and others are avoided. There are future riches to contemplate, and possible early retirement, or semi-retirement.

In contrast though, business has to be won, and many sole traders report the number of working hours as being excessive, especially in the early days of a full-time venture. This includes working through illness, without sick pay and the other relative comforts afforded to salaried staff. Family/partners need to be supportive of the venture, and be comfortable with potential financial pressures, reduced time for family activities, etc.

Before self-employment is contemplated, a rigorous evaluation of all risks and opportunities as well as emotional and lifestyle factors is needed. This appears to be obvious advice, but there are small businesses that fail because they are inspired primarily by enthusiasm, and lack the necessary market research and evaluation of their true prospects

The highs and lows of self-employment

Self-employment has many drawbacks as well as opportunities and perceived lifestyle advantages. Self-employed surveyors report the highs being higher and the lows being lower than with regular employment. Frustrations include long periods with relatively little work to undertake (leading to the viability of the venture having to be reassessed), and clients being slow to pay, not forthcoming with approvals to recommendations that enables casework to be progressed, and unable to offer further work. Pitches for new work take time, and prospective clients' criteria and decision-making processes to appoint consultants is at times fickle.

A partnership enables contact to be maintained with other people, but for a sole trader with few meetings, inspections, etc, to undertake, many working days are spent alone. If finances become squeezed, and pressure is applied by lenders, the family home might need to be re-mortgaged — or even sold in order to move to a smaller house and remain solvent.

Commitment, resilience, determination, motivation, patience and toughness are key qualities required if experiencing such difficulties. Building a business takes time, and many people never take the step. A common reason is the fear of failure and lack of confidence (if not ability). Most businesses suffer difficulties, but provided the underlying concept is strong, and the cash position of the business not too precarious, they can survive them.

One telephone call from a prospective new client offering an instruction could see the fortunes of a small business change, and the spirits of the surveyor uplifted. Previous business development initiatives, the success of which was impossible to measure at the time with any accuracy, all of a sudden inspire a lead. Clients introduce other clients, and once a job has been done well, others follow. New clients emerge through unpredictable sources, and good fortune also helps achieve vital breaks.

In time, a regular client base emerges, and a steady level of work eases the early pressures of self-employment. Expansion of the business, and the taking on of other staff, might not appeal, but many sole traders enjoy years of self-employment, combining business and leisure interests with immense satisfaction.

Scope to work from home

Traditionally, surveyors setting up their own practice have often acquired office space and taken on a secretary. Increasingly, it is feasible to work from home, drawing on the latest information technology, without secretarial support. Benefit is gained from a lower cost base which facilitates more competitive pricing, higher profitability, and less overall risk.

For example, a residential agency needs an office/retail outlet, and would have to employ secretarial/administrative staff, whereas a surveyor providing specialist consultancy advice to clients on a national basis typically works from home — and if proficient with information technology, does not need a secretary.

The impression given to clients by working from home needs to be considered. Meetings at clients' offices or other venues helps avoid problems, and the location and size of a home could be impressive. Background noise, however, including from young children, gives a poor impression when speaking on the telephone or conducting meetings at home.

Home-based sole traders often report that it takes time before they are able to separate home life from work and therefore relax at home — especially if working long hours and if keen to develop the business. A small office elsewhere helps avoid this, even if it is not essential. An alternative is to share facilities with another business, making dual use of a secretary, administrative support, etc. One benefit of a home-based business is that it provides a low-cost start to a venture, with a view to expanding into offices when sufficient instructions are won.

If taking on premises, surveyors are naturally well-placed to deal with the property issues, and guidance is not provided in *Starting and Developing a Surveying Business*. If need be, the free RICS publication, *Property Solutions — A Practical Guide for Your Business* is available. This summarises the issues for businesses to consider in respect of property. Office costs are influenced by location, style, size, and the impression that premises need to convey to clients/customers. Surveyors should be aware of the need for buildings insurance, contents insurance and public liability insurance, including insurance for loss of fees in the event of fire/destruction. Home insurers need to be contacted to see if business use makes a difference to the policy and premiums paid. Car insurance similarly needs to include business use if not already doing so.

In theory, a liability for capital gains tax, business rates, etc, could arise if working from home, but in practice, the nature of a sole trader's business use does not cause problems (ie by generally using shared domestic and business space rather than designated business space). If however, staff begin to work from the surveyors' home, and dedicated offices are created, problems are more likely. Planning permission could be required to operate a business from home, and lease consents are likely to be needed if the property is rented.

Necessity of secretarial support

Sole traders who are reasonably fluent with information technology usually find that a secretary is not necessary. Optimum use should be made of a mobile phone, laptop and other devices. Even if it is important that the telephone is answered when in meetings or away on holiday, calls can be diverted to a designated mobile phone of a family member or friend who is able to respond professionally with the name of the surveyor or company. There are also companies who provide secretarial/telephonist support services.

The ability to type at a reasonable speed directly into a laptop or desk computer and prepare letters and reports is considerably more efficient than the traditional form of dictating letters. Where surveyors lack IT fluency, and have relied heavily on secretarial and administrative support in previous salaried employment, they need to consider basic IT skills training before starting their business. Similarly, they will need to be appraised of the latest IT available, and how it is operated without the day-to-day support of others.

Sole traders are also often able to draw on family members, friends and neighbours to assist with photocopying, banking, posting letters, buying stationary and many other small roles — all which would otherwise add up as non-chargeable time for the surveyor. Time is money for the sole trader, and getting the many small elements and overall logistics right accumulate to having a significant effect on end-year profitability.

Secretarial services need not be full-time. Options include informal arrangements with family and friends, sharing secretarial facilities with another surveyor or local business, using a recruitment agency, and employing a secretary part-time.

Research into the options and costs of secretarial support, IT needs, and a car purchase, are just a few examples of the early tasks which must be undertaken when setting up as a sole trader. As a general example, costs are on the lines of computer, software, telephone, fax, mobile phone, e-mail/web facility, chairs, desks, cupboards, bookcase, shelves, filing cabinets, camera, safe, pictures/

decoration, plants, reception area, insurances, vehicle, premises costs (rent, rates, water, electric, gas, service charge, repairs, etc), telephone charges, other equipment, staff, stationary, postage, travel/mileage, advertising/marketing, lawyers/accountants/other professional fees, professional indemnity insurance, subscriptions, journals, etc.

Profit and profit margins

Surveyors, as with many other professional services businesses, commonly focus on the fee levels achievable for particular services when determining the nature of their business, and the surveying activities to be undertaken. However, cost structures and profit margins differ between surveying businesses, services offered, and working methods. Surveyors need to assess profit and profit margins, not only fees.

As a simple illustration of the interrelationship between fees, costs and profit margins:

> 30 chargeable hours per week, for 50 weeks, at fees of £100 per hour, working to a profit margin of 25% (reflecting office, staff and other overheads in a larger practice), produces a profit of £37,500 (£150,000 fees/turnover, £112,500 costs, with the 25% profit margin deriving from £37,500 ÷ £150,000).
>
> At profit margins of 75% (reflecting home working, no other staff and minimal overheads for an individual surveyor), fees of just under £67 per hour for the same chargeable hours see profits double to £75,000 (£100,000 fees, £25,000 costs, with profit margin calculated as £75,000 ÷ £100,000).

This shows how an understanding of the markets for surveying services, and the trade-offs between services, fee levels and essential costs, helps a particularly profitably venture to be established, even when, on first appearances, fee levels seem low, and profitability is therefore thought to be limited. For surveyors having worked previously in the larger surveying practices, the illustrations show how the economics and opportunities for small new surveying businesses are very different to the costs, margins, and business models with which they are more familiar.

There are, of course, many factors to consider when determining the format for the business, but this shows how with a skilfully tailored concept, a busy and reasonably talented home-based self-employed surveyor is able to achieve strong profitability. As an example, in the late 1990s a Midlands based surveyor in his early 30s, starting a business from home without secretarial support or existing contacts/clients, reached annual profitability of £100,000 within the first full financial/tax year. A combination of property investment advice, property management work and business management consultancy enabled him to provide high added value to growing small investment companies and other businesses at affordable fees. A low cost base was secured through home working, and basic information technology skills avoided the need for a secretary. The surveyor's equivalent salary if working for a property consultancy would not have exceeded £30,000. Today, the surveyor enjoys a secure financial position and need not work again, although he still combines property, advisory and educational interests. The principles embraced by the surveyor are reflected in the issues emphasised in *Starting and Developing a Surveying Business*.

Fixed costs, variable costs and the surveyors' commitment

A sole trader is likely to have a relatively high proportion of fixed costs compared to variable costs (as can be seen from the list on p 4) — although the cost base depends on the type of work. As a simple example, a self-employed surveyor undertaking 40 hours per week (30 of which are chargeable time), may achieve £30,000 profits from £45,000 income and £15,000 costs. A further 10 hours' chargeable work (two hours per day) would increase income to £60,000, and produce profitability of £45,000. In other words, a third extra chargeable time achieves 50% increased profit. This assumes that all costs are fixed, which except for minor additional costs is not too unrealistic for a sole trader — with office equipment, the cost of acquiring a car, computer and other IT needs, the RICS subscription, *Estates Gazette* subscription, etc, remaining unchanged. Further detail on costs is shown in Chapter 8 and Chapter 11.

This is similar to the enhanced profitability achievable when property consultancies push their surveyors to exceed 40 hours per week. Although bonuses are usually payable, and are a variable cost to the company, the surveyor's extra efforts have a disproportionately beneficial effect on end year profitability, especially if profit margins are in the region of 10–20%. At 15% profit margins, for example, (£100m turnover, £85m costs, £15m profit), a 5% increase in turnover increases profit by 33% to £20m (assuming no cost increases). Conversely, lost fee earning time quickly erodes profitability. The extent of such sensitivities depend on the typical margins of the business, and also the relationship between fixed costs and variable costs. The smaller the margins, the greater the sensitivity of variations in income and costs to profitability. Likewise, profits are more sensitive to changes in income when there is a higher proportion of fixed costs. In the above example, the sole trader is initially working to margins of 66% (£30,000 ÷ £45,000). Larger practices see profitability vary, but between 10-20% as a guide, depending on the nature of the business, and the performance in a particular year. Smaller niche firms, but with turnover still in excess of £1m, achieve margins as high as 30%, and occasionally more. As a general point, margins for a sole trader taking all profits and larger practices having staff costs are not really comparable — but they still illustrate certain advantages enjoyed by the sole trader.

The above illustrations show the scope for sole traders to increase profitability by working hard (such as two hours extra per day increasing profits by 50%). The same effects derive from productivity and efficiency, with one surveyor cutting through work more quickly than another. The above points regarding profit/profit margins, fixed and variable costs and the surveyor's commitment are important when considering the services to be provided (see below) and the working methods adopted (see Chapter 2 for more detail).

Skills review

The services which a surveyor is able to provide depend largely on previous experience gained in practice. The surveyor needs to review his existing skills (which are likely to represent the strengths of the new business), and also others which could be developed in order to provide a wider range of services.

For example, a surveyor with recent experience only in retail agency had still maintained a working knowledge of landlord and tenant. This was achieved through negotiating lease terms which reflected the need to enhance landlord clients' investment profile and estate management requirements, and through scrutinising assignment and sub-letting provisions so that tenant clients could minimise their property liabilities. He had not undertaken rent reviews and lease renewals for many years, but committed study and discussions with colleagues and contacts quickly brought the surveyor up to date. The level of expertise required by the surveyor depended, of course, on the type of clients sought, and the level of work being undertaken. Alongside agency work, it was possible to undertake landlord and tenant work in respect of smaller properties, often against unrepresented opponents, as part of a range of general practice work. It was not feasible to seek work in respect of larger property interests, where an up to date knowledge of case law, and an ability to take cases to third party would be essential. As part of the business plan, however, landlord and tenant expertise was to be developed through further suitable learning activity. This is a good illustration of how lifelong learning/CPD impacts directly on business success.

Securing an edge

Sole traders should evaluate their strengths and weaknesses, and establish the scope to secure an edge in their potential markets. As another example, a sole trader living in north London, with many years' experience in planning (including a high level of expertise in environmental issues and contamination) aimed to be viewed by clients as a better option than the larger practices. He knew that some clients of the larger practices take the view that despite initial pitches for business and subsequent client contact being led by senior personnel within the practice, certain day-to-day casework is quickly passed to graduates and others who are unlikely to deploy the same expertise — with fees still exceeding those of the experienced sole trader. The sole trader's own expertise, and offer of personal service, was thought to present an edge as part of the business development strategy. He also recognised that the expertise of key individuals in a large practice, and the wider range of services on which the large practice draws internally, makes it difficult for the sole trader to win instructions. Clients sometimes feel more reassured with a larger company, even if only to minimise their representative's scope to be blamed personally if anything goes wrong. Also, a self-employed surveyor represents a risk to a client if becoming too busy, or suffering serious illness, whereas larger practices are better able to manage such difficulties and ensure continuity of service.

The uncertainty as to the scope to win work from clients typically appointing larger practices meant that the surveyor's business plan was to target both higher level work from national clients (environmental and contamination expertise), and lower level work from local clients (mainstream planning work). This enabled a start to be made to the business, with the desired client base and superior fee levels being established more progressively. Also, as the surveyor wished to expand the business, another surveyor could in due course be taken on to manage the straight-forward, locally based, casework, while the principal surveyor concentrated on the development of the more lucrative national client base.

Track record and profile are key qualities for the sole trader in securing clients who usually appoint the larger practices ahead of smaller practices and sole traders. An ability to win business on the back of both "price" and "profile" is a key determinant of success and profitability at all levels within the property consultancy market. An efficient cost base is also, of course, important.

Market analysis

An analysis should be made of the current market structure for the proposed services. *Estates Gazette Directory* (the monthly supplement within *Estates Gazette*) shows the vast number of surveying firms in London (among other locations) — although there is clearly also a large potential client base. As well as the main international private practices, there are mid-sized firms operating only in the UK, and some predominantly in London. There are smaller firms, some covering only parts of London, and there are sole traders and small partnerships. The international practices provide the usual full-range of mainstream surveying services, but some are stronger than others in certain areas, including specialist fields in which competitors have little, if any, expertise. Small and mid-sized firms specialise in retail agency, or development work, for example.

The sole trader setting up in such a market often looks to create a niche/specialist service, such as "contamination and environmental specialist", "rating specialist", a "dilapidations expert", or "landlord and tenant surveyor". An analysis needs to be undertaken of fee levels among any competing sole trader specialists, and within the larger firms. Surveyors will be familiar with the typical bases for setting fees, but as part of their market research, they should still establish the fee levels achievable as a small firm from typically smaller clients. Although rates per hour are adopted in the earlier illustrations, fee bases other than a standard rate per hour will apply, such as percentage of rents/sale prices for agency, fixed fees, and incentivised fee structures (such as for rent reviews, lease renewals and rating). In practice, a surveying business experiences highly profitable instructions (such as quickly concluded lettings, sales and rent reviews), and instructions which yield nothing (such as failed agency instructions without an abortive fee). Further comments on fees are included in Chapter 6, Winning Business.

Specialist services

Specialist services generally give scope to command higher fee levels than a range of general practice work, and also enable clients to be more easily targeted as part of business development initiatives. Bigger clients tend to be more used to paying higher fees to large practice advisers. However, specialist skills typically involve covering a wider geographical area, incurring greater travelling time and additional costs.

It is often more difficult to expand a small business offering specialist services — not just because the market is smaller, but because it is harder to find staff or partners with the necessary expertise, or interest in joining a small firm. For example, a surveyor setting up as a sole trader in the north west providing a range of general practice services was quickly able to expand a new practice by recruiting friends and old colleagues from both private practice and the public sector, and taking on a graduate. In contrast, a self-employed London based surveyor specialising in out-of-town retail development consultancy, felt unable to expand because of the difficulty in finding someone suitable, and the additional risk of a further surveyor departing, thus affecting the continuity of the business.

Where the self-employed surveyor provides specialist services to larger clients, there tends to be a greater expectation that the surveyor will handle the work personally. This generally makes it harder to introduce new personnel/surveyors to clients, compared, for example, with a small general practice consultancy with smaller businesses as clients, who are attracted to the firm and its services rather than specific individuals. As shown in Chapter 7, Business Growth, and reinforcing the above illustration regarding "profit and profit margins", many sole traders find it preferable not to expand, and wish to remain as the only surveyor in the business. Exceptions include surveyors looking to grow into a large regional and possibly national practice, rather than to a size of say, two or three, surveyors which serve a local market.

Opportunities in crowded markets

Although the ability to exploit gaps in a market helps make a business successful, it is still possible to do well in a crowded market, particularly with a low cost base, aggressive business development initiatives, and good quality services. Indeed, entering a larger crowded market with the right expertise would generally be more profitable than seeking to exploit a perceived gap in a market which is too small, or is not conducive to winning businesses easily or cost effectively.

In a crowded market it is particularly important to consider how services are differentiated from those of competitors. As a simple example from the above illustration, a more precise description of "Contamination and environmental specialists" would be better than "Planning and development advisers" particularly if targeting the higher profile clients in respect of contamination and environmental work. Two distinct trading names could be adopted: one for the national market and one for the local market.

Level of market entry

Surveyors entering a market at a lower level than their capability suggests is appropriate, should gain a good start to the business. The fee levels are lower, but such an approach helps a business find its feet, and its principal/s to gain confidence. Better quality work and higher fees/margins can be achieved in due course — although as indicated above, apparently lower value work with lower fees could still prove to be the most profitable if associated costs are far lower. If young surveyors wish to develop their own business, they could appear too inexperienced to clients, and struggle to secure work, even if they are capable of superior performance to more established competitors. Beginning at a lower level provides a better chance of the business being established successfully.

One route to substantial business success pursued by relatively few people is indeed to enter a market which is beneath the level of experience, expertise and talent they possess. An edge is quickly gained on competitors, and a position as market leader soon secured. It takes an entrepreneurial eye to see and exploit the right concept — as indeed it does in formulating the best blend of services and working methods for any new surveying business. For example, around five years ago a chartered surveyor with previous experience only in commercial property established a residential estate agency in a large market town. It now leads the local market. The business draws on the "chartered surveyor" profile, utilises imaginative marketing and secures an all-round edge on relatively complacent, often long established, competitors who have each lost a share of the market to the new firm.

Although it is suggested that people need a challenge in their work, and that work which is too basic is not stimulating, the sole trader's challenge of running and also developing a business tends to displace the need for a more intellectually rigorous workload. The scope for all work to contribute to profitability and personal financial rewards is also more enthralling than working to a standard salary, albeit possibly plus bonus.

Benefit of business acumen

As a general point, there are usually good business opportunities in markets whose participants generally lack certain business and managerial skills — including an eye for new markets, differentiated markets and niche markets, general areas of business to exploit, and an instinctive ability to read markets. This is one of the reasons why surveyors with the right qualities gain considerable financial rewards in the property industry (with RICS Agenda for Change in the late 1990s, for example, highlighting the general deficiency in business and managerial skills among surveyors).

How business will be won

Other factors determining the services to be provided include how business is to be won for particular services, who typical clients will be, and the appointment criteria adopted by prospective clients. Public bodies and larger commercial concerns/clients, for example, will usually tender instructions. Surveyors might need to be on a panel to even get the chance to tender. Without the right profile and track record, it is difficult to win instructions in such circumstances, and it is easier to start the business with smaller commercial concerns/clients. Commercial clients need not always tender to a number of surveyors/practices if the services on offer appear sufficiently attractive and cost competitive — and if the surveyor performs well, periodic re-tendering is less likely.

In the early stages of the business, it is important to generate momentum through early successes. Profitability should not be an undue early concern — although from a psychological perspective it could be the only tangible measure of success. Vision and belief is required. In due course things usually click and one job leads to another.

One-off and ongoing instructions

Certain lines of business relate to one-off instructions, and each job effectively incorporates the time and cost of business development activity, including expenditure on advertising. Examples include single sales or acquisitions for small clients, a rating appeal or a rent review. Further work could, however, be available five years later for a rating appeal or rent review: one example of business development initiatives being to diarise the client's next likely requirement, and make an early approach to win the next instruction. Other work is generated more regularly from key clients, and instructions more often lead to further work (including from parties other than the client — such as one rent review picking up instructions from neighbouring tenants/new clients). All such factors affect the extent of chargeable time, the pricing of services, costs and profitability.

Finding good clients, rejecting bad clients

Some clients enjoy "playing the client", putting their advisors on edge, assuming that the market is hungry for their business, and looking to cut fees where possible. Advisers are not usually as desperate as such clients like to think, and certain clients will be avoided altogether, or their work afforded relatively little priority. As self-employment develops, the surveyor is more content not to work for the poor payers and those clients causing other frustrations. Ideally from the outset, surveyors are not prepared to work at any price for any client, and instead have confidence in their vision for the venture, and await better opportunities.

A good client understands the cash flow situation typically faced by a self-employed surveyor, and makes a conscious effort to suggest that invoices are submitted, and to process payment quickly. Similarly, from the sole trader's perspective, another key (often underestimated) aspect of running a successful business is "being a good client" — and getting the best out of professional advisers such as accountants, lawyers and business consultants, really in the same way that good employees need motivating and rewarding (see "Advisers and suppliers" on p 16).

Rolling out services

Entrepreneurial elements in business include establishing services which can be rolled out in a way that sees further fees derive for relatively little additional work. This is not as achievable with consultancy services, as with sales/retailing, for example, where identical items are produced, but opportunities exist. Once research is undertaken and particular knowledge gained from certain case work, further similar instructions are undertaken in considerably less time, and therefore at higher profitability than competitors having to undertake research and gain experience afresh. The right concentration and/or combination of services therefore adds to profitability. Efficiencies are also gained by having clients close to one another, and co-ordinating meetings and site visits on the same day. Examples of other fee lines include introducing business to other agents, and commissions if arranging insurance.

Rolling out a concept also applies to expansion through further offices in new locations (as demonstrated in the example in Chapter 8). Here, experience gained in setting up the initial office enables further offices/business to be developed with greater efficiency and confidence, and therefore enhanced profitability.

Cash flow, and defensiveness of fee lines

Certain lines of surveying work take time before fees are billed (such as a drawn out rent review or rating appeal), whereas others see instructions completed in a matter of weeks, and invoices immediately issued (such as valuation). Property management work, and the collection of rents and service charges, provide regular income, including through the authorised deduction of fees.

The level of agency work is influenced by the state of the economy and property market — and if deals are not achieved, fees are not likely to be received. The performance of property markets is also localised, with a particular town or city seeing either a considerably higher or lower level of activity than the national trend — thus creating both opportunities and risks for the small surveying business exposed primarily to the local market.

Rent reviews and lease renewals usually need to be undertaken when due, but if, for example, the market has fallen over the past five years, and rents cannot be increased, rent review work dries up, and landlords do not seek to progress all lease renewals. The surveyor therefore needs to be alert to such trends, and again be looking for opportunities, but also seeking to manage risk — such as looking for rent review and lease renewal instructions in advance from clients in a strong market, or seeking to concentrate on other services if such work is likely to be limited.

The level of rating work is likely to vary, depending on the timing of rating revaluations, and also on the extent to which the government/Valuation Office Agency pursue an approach of being "right first time", and thus limiting the scope for appeals. Although there are many aspects of rating other than simply reducing a rateable value, a successful "right first time" approach could severely dent the profitability of a small rating practice. This is also a good illustration of the benefit of a balance of services — and/or the ability for a business to develop different areas of expertise within its markets if need be.

If a small practice includes residential estate agency, an increase in interest rates could dampen activity in the market, and see profitability fall sharply. Again, a diversity of activities minimises such effects.

The rise and fall of certain sectors also sees a surveyor's work diminish. In the telecoms sector in the early 2000s the telecoms operators' need for property acquisitions to accommodate telecoms equipment and secure coverage for their networks helped provide strong earnings for those surveyors/practices having entered the market in anticipation (with a business plan recognising the likely short term potential of the market). Operators' and therefore surveyors' acquisition activity inevitably later subsided, and although increased telecoms property holdings created more property management work, more surveyors/practices had entered the sector. When the commercial imperatives of the operators were to roll out their networks, fees paid to property consultancies were a relatively small consideration — especially as a consultant's expertise and profile could influence whether a planning consent was won or lost for a site. But once the operators began to manage an estate comprising numerous low rental interests, in a more competitive market for surveyors/property consultancies with expertise in telecoms, the profitability to the surveyor/property consultancy became more limited. Potential instructions also diminish if an operator, like any large client, decides to undertake work in-house through an expanded property team.

It is not common for surveyors to seek to get in and out of a particular market, and "make their money and run" in the same way that retail and leisure operators exploit fashions, trends, etc, but the astute surveyor still searches for such opportunities. As well as telecoms, access audits pursuant to Part III of the Disability Discrimination Act 1995 (effective from October 2004) and asbestos regulations (effective from May 2004) have provided surges of profitable work for those surveyors/practices suitably positioning themselves in advance. This is, however, a good illustration of a market where individuals could make an apparently good decision to develop new fee lines, but the aggregate effect is one of over-supply of surveyors' services, and work being spread round too thinly at lower fee levels than initially anticipated.

The above illustrations show that for a small surveying business, a variety of work helps cash flow, and eases risk. It also helps clients to receive the majority of property advice from one surveyor/practice — minimising the scope for clients to benchmark advisers, and enabling the surveyor to pitch fees with less fear of competitive threat. It is also important though to not be seen as a "jack of all trades" surveyor/practice, covering such a wide range of work that clients question whether one surveyor is able to do all the tasks justice.

Securing a mix of clients

A self-employed surveyor could secure a mixture of work from many clients, or a high level of work could be generated from only a small number of clients. A large amount of work from one client provides a welcome start to self-employment, but it is important to secure other clients in due course — and not be dependent on the level of work provided by one or two key clients. Where self-employment involves only one client, such as a previous employer, this is unlikely constitute self-employment in terms of taxation, and as indicated in Chapter 5, accountant's advice should be taken.

Thought should always be given to the likely level of work being undertaken three to six months ahead, as well as to the longer term plans for the business. Particularly if the business is involved in project type work, or larger requirements for certain clients, it could go from being very busy, to having relatively few instructions. Marketing, business development, and contact with clients who are not currently providing instructions, is therefore just as important when the business is busy, as when work appears to be drying up.

Complexity and diversity of work

Another element in determining the services to be provided is that more complicated, albeit higher fee, work may take longer to progress, and be less profitable, than lower fee work which is capable of being completed very quickly. Intense, complex work is also more tiring, and straightforward work enables more hours to be committed. This is another reason why a diversity of activities is helpful.

A balance needs be secured between areas of practice operating to pressurised timescales (such as a valuations often being required within a week), and those undertaken in a relatively leisurely way (such as the negotiation of a rent review — although involvement in arbitration usually involves a large amount of work in a short space of time). The ability to undertake certain urgent work if required by clients is a possible selling point for the surveyor. In the case of the Midlands based surveyor referred to earlier in the chapter, evening and weekend availability was attractive to clients who valued the impression of intense loyalty and commitment.

Different areas of work also involve a different level of telephone calls, meetings, ease of using e-mail, the need to be instantly contactable, etc. The self-employed surveyor should analyse all such factors in developing finely tuned methods of working.

As indicated in the next chapter, a diversity of work helps increase motivation, commitment and profitability. As well as undertaking mainstream surveying work for clients, self-employment could extend to personal property investment and development interests, university lecturing and also non-property ventures.

Issues with existing employment

If a surveyor is to begin a business while already in salaried employment, the present employment contract needs to be checked, as this might limit the work undertaken, the clients that are taken on, the geographical area covered, etc, once having left the firm. Advice should be taken from a lawyer as to the validity of any such contract terms.

The surveyor could still negotiate suitable other terms with their employer if looking to leave and set up their own business. An employer is likely, for example, to welcome a longer notice period in order to provide greater continuity in handing over work to colleagues. On other occasions, the surveyor is immediately placed on "gardening leave" by the company — providing beneficial time to set-up a new business, if not necessarily commence trading.

A relatively straightforward option, which takes out many of the pressures in developing a small firm, is to begin with a number of clients that are sourced from the current employer. This is not to suggest ethical impropriety, as this could be undertaken in agreement with the employer as part of ongoing working relationships. Here, good relationships with clients, and a strong profile and track record among the surveyor/s involved, makes the new practice an attractive option to clients. Clients must however be comfortable working with a smaller consultancy. For areas of surveying such as valuation for lending purposes, this is not usually feasible, as lenders require larger firms on their panels. Contacts, family connections, etc, also help an easier start to be gained to the business. Other benefits for a business fortunate to enjoy such an immediate client base include reduced costs such as on marketing, fee income being earned sooner, cash flow being less problematic and finance requirements being lower. The ultimate satisfaction for sole traders though is achieved when beginning from scratch.

Other issues with partnerships

In some partnerships, partners have different areas of expertise and are responsible for different areas of the business, whereas in others, all the partners undertake similar work, perhaps with designated responsibility for looking after particular clients. A partnership is more likely to require its own offices, and be a larger scale operation compared with a surveyor setting up as a sole trader. Individual partners could, however, work from home, although in the early stages it is important to spend time together in generally establishing the direction of the business.

The partners need to be aware of the work in which each other are involved. One partner's work for a client might lead to work that another partner could undertake, but also, partners wish to see each other committed to the business, and not spending time on other interests. Arrangements in respect of the share of profits ease such difficulties, but conflicts can emerge — such as when one of the partners is highly motivated and anxious to develop the business, and another sees it more as leading to semi-retirement. Partners setting up a business usually know each other, and there should be trust. However, friends disagree, and relationships break down, especially when a business is struggling.

Peoples' personalities and attitudes also change when their professional life begins to see substantially increased financial returns. Twenty to 30 years of hard work at a reasonably modest salary carries a certain lifestyle, and once annual income rises considerably, some personalities chase the lifestyle to which they had always aspired. One partner might develop a hunger for more money, and wish to work ever-harder, while another seeks increased leisure time. Business people seem to be great friends when things are going well, but difficulties in a business see individuals sidelined. This also influences family and friends if social activity has been built around professional interests. Individual partners/principals always need to guard their backs within the business, as well as maintaining an alertness to external threats to the business and themselves. As a general point, apart from "money does not bring happiness" being a consoling phrase for people with none, research studies have shown that this is generally true, and that money instead brings new anxieties (and widens the scope for conflicts with business partners).

A surveyor working as a sole trader has the option of developing the business by taking on a partner — as has a partnership of taking on further a partner or a limited company recruiting a surveyor who takes a position as director and/or shareholder. Consideration has to be given to whether an initial payment is required from a new partner, or whether there are mutual advantages which will be reflected in enhanced profitability and any other benefits. A sole trader or partnership which has become successfully established does, of course, have a client base and instant revenue stream from which a new partner should benefit.

Advisers and suppliers

As mentioned above, in the same way that people need motivating, the right attitude needs to be afforded to suppliers/professional advisers. Surveyors will be aware how highly valued clients, who pay good fees, are afforded greater priority than those generally seeking to negotiate fees downwards, and do not provide much work — including spreading a lot of work too thinly among a range of property advisers. Consideration always has to be given to how to get the best out of professional advisers. Lowest price could mean lowest quality, and some advisers are even "reassuringly expensive".

With advisers such as accountants and solicitors, regard has to given to how assiduously they reflect the surveyor's circumstances as opposed to working to broad principles. This is similar to surveyors' work, such as a rent review: does a surveyor fully address a client's issues and objectives and take a tactical approach of delay while better comparable evidence emerges in order that the highest possible rent can be secured, capital value uplifted and greater profits realised on sale — or open negotiations immediately, with a view to quickly completing the case and billing for fees sooner.

Some advisers are also too risk averse, and frustratingly for the client, fail to take a sufficiently commercial view — nor have the appetite for a commercial dispute, or an argument with the Inland Revenue or Customs and Excise. Another unique selling point of a sole trader might be the tenacity, aggression and loyalty devoted to client affairs, and which is perceived by clients to be unrivalled by salaried surveyors in competitor firms.

Successful business people

As with all business people, the surveyors who generally do best are those with an entrepreneurial eye. They instinctively see gaps in the market, and understand how sub-markets and niche concepts can be exploited. A combination of technical knowledge, expertise in practice, an ability to read markets and basic business skills creates an edge on others. Incisive thought processes and the ability to carve out sharply focused concepts see strong demand for the services on offer — the profile and exclusivity of which cannot be matched elsewhere. Intelligence is not the same as thinking deeply, analytically, creatively, and with vision — nor being able to think quickly and correctly, especially when under pressure.

The instincts of highly successful business people are hard to explain, and difficult for others to understand. They just "seem to know", and their strategies reap success. While the masses follow the herd, successful business people focus on ways to add value to existing services/ventures, or create fresh themes. Instinctively, successful business people see how relationships between revenue, costs, margins, etc, combine.

Psychology is important in business: such as how to captivate consumers/clients, how different clients make decisions to appoint advisers, what motivates staff and how the market perceives a business/brand. Confidence is also a key element in driving forward a business, just as it is in the performance of economic and property markets. The building of relationships, use of contacts and ensuring trust among all those with whom business is conducted is important.

Surveyors should draw on the traits of highly successful business people, and apply elements to their own business. As indicated above, high financial rewards are achievable, but that does not belie the difficulty in formulating the right concept, and being able to win sufficient business at the right price.

To a large extent, the laws of supply and demand determine the viability of setting up in business. In certain locations, and with certain areas of existing expertise, the venture will simply be unviable. However, as mentioned above, one surveyor does better entering an already crowded market because of particular initiatives deployed, than another surveyor with niche services of great potential, but not the necessary skills to win business or manage the venture effectively. Similarly, the assessment of the interrelationship between certain services, fees, working methods, costs, margins and profits (for both the business as a whole, and individual fee lines) sees highly profitable ventures established, where others cannot see such dynamics.

Consideration is given in the next chapter to personal qualities and working methods, and Chapter 3 examines financial issues. Chapter 4 on business plans includes further points in respect of the direction of the business, and management issues. Chapter 7, Business Growth, introduces issues associated with expansion, and which should also be drawn on when considering the format initially to be taken for the business. Chapter 6, Winning Business, is relevant to early considerations, as the proposed services need to be conducive to cost-effective marketing and the securing of clients.

Examples of new ventures

The differing circumstances and opportunities for surveyors to establish themselves as sole traders are shown by the following examples (which also includes one example of a partnership).

General practice and specialist services — city:
Anthony Gallagher

Previously having worked for a large city practice, and becoming tired of the monotony of the same journey to work, colleagues and workload, Anthony, at 53, left an approximate £50,000 salary plus other benefits, to set up alone. Although a specialist in landlord and tenant work in his previous employment, he needed to offer a wider range of services to clients of the new business. He decided that initially the business would span larger clients (such as property/investment companies) in respect of landlord and tenant work, and smaller clients regarding other general practice work (including valuation, rating, lettings and sales). He established an office in a high street suburb near to home in order to help raise the profile in the immediate locality, and to show the stature of the venture to all prospective clients (including regional and national property/investment companies in respect of landlord and tenant work). For non-landlord and tenant work, the practice had a large potential market throughout the city, but agency work had to be confined locally owing to the time incurred in travelling for viewings. In contrast, landlord and tenant work could cover a relatively wide geographical area, as inspections and meetings would be relatively limited, and fee levels sufficiently high. While agency was not thought to be particularly profitable in the small local market, the presence through site boards and advertisements in the local newspaper helped secure instructions for other work. A niece provided secretarial and administrative support, and start up costs were financed easily from savings, with a home clear of mortgage commitments providing security as a fallback option for raising finance if need be. Anthony's intention is to work alone through to retirement at around 65, hoping to sell the business, but retaining one to two days per week part-time involvement. Such has been the level of work established to date, expansion is in the process of evaluation.

General practice — rural/county: James Shelbourne

Having gained 11 years' experience working in a public sector property team and six years' experience in a corporate sector property team dealing mainly with retail property, James, at 39, established his own practice following redundancy owing to outsourcing. The rural/county location provided a limited immediate market, and the business needed to cover a geographical area of up to 75 miles from James' small home town. 45 miles north, in a larger town, was an established practice of commercial property consultants who had grown over the last 20 years to three partners, three qualified surveyors and four support staff. Although there were several other self-employed surveyors regarded as competitors, the established practice regularly won much of the higher value work in the county, and in parts of neighbouring counties. In view of the geographical spread of clients and in order to secure a low cost base, James established the practice from home. The address details stated on letterheads and other stationery ensured that this was not obvious to potential new clients, and meetings were usually arranged elsewhere. James considered adding on to stationery that the business also had offices in two other towns in the county (drawing on family addresses), in order to appear to be a larger firm than was actually the case. It was felt that this would be misleading, and would not create a good impression when clients were aware that the business was run by one surveyor. If, however, the business expanded, and a partner or other salaried surveyors were taken on, other offices could be used in order to convey the county-wide nature and size of the business. This would also help win business from clients who considered the agent's close proximity to infer local expertise and good service (and which is always a good selling point when seeking new clients).

Having not worked in private practice previously, extensive preparation for the business was needed, including through liaison with old colleagues who had since joined private practices, albeit away from James's home area. It was thought helpful to seek early instructions in respect of lower value properties, and generally ease into the new environment. This was helped by clients typically having little or no property knowledge. Research was often, legitimately, undertaken as certain case issues arose — such as the rights of a residential tenant, and aspects of rural practice. It was, however, important from the perspective of reputation as well as professional ethics that the surveyor did not take on instructions without the required level of expertise — noting that negotiations would be undertaken with other surveyors. Nevertheless, it was accepted that in more remote locations, the local surveyor more often, and legitimately, tends to become involved in areas of work despite their previously limited experience. Clients' main alternative is likely to be the instruction of a larger firm, located many miles away — and whose fee levels are unlikely to be acceptable to the smaller local client. This was the case with the established practice in the county, and James slowly began to win more instructions. There has not, however, been a high volume of work. James is currently seeking to develop other skills, including a more detailed knowledge of rural property issues, and also compulsory purchase (building on public sector experience in the 1980s). This is because former colleagues, now in a large practice, are increasingly involved in planning and regeneration work, and lack expertise in compulsory purchase. Therefore, as well as local clients, James will provide consultancy services to the practice. Once new skills and an expanded client base are settled, James is keen to consider ways to develop the business, and be recognised as a leading consultant in the county who is able win business typically won by the larger established practice.

Asset and investment management, London West End — three senior surveyors as partnership: Donald Copley, Maria Sergeant, Eric Davidson

In the city of London, three colleagues had reached their 50s — Donald and Maria with the same private practice, and Eric working with another practice. Their various children had been seen through university, current salary and bonus levels were comfortable, pension arrangements were in place, mortgages cleared, a comfortable cushion of wealth assured, and respective positions in local social scenes provided healthy interests beyond work.

However, Donald and Maria each needed a new challenge that their employer could not provide owing to the constraints enforced by group company issues, and increased infighting among fellow senior personnel. As well as their expertise in portfolio management, investment agency/acquisition, etc, an expert in property management (landlord and tenant, estate management, asset management, etc) was needed, and Eric was pleased to be approached.

It was thought that the combination of investment and property management work in which the three colleagues were involved on behalf of clients reflected the stature of the present employer, and that it could be difficult, as individuals, to develop a small practice needing work from large clients. Nevertheless, offices were secured on a favourable sub-letting arrangement, and prospective clients became aware of the partnership following a series of business development initiatives. The partners sold their individual track records, their contacts in the investment world and their expertise to clients. Another attractive feature was their ability to offer competitive fees due to low overheads. The level of personal service was stressed, together with examples of how the surveyors' niche expertise would add financial value to clients' property interests. All went well, and although the initial few years' remuneration did not match that of previous employment, the practice was able to slowly expand. A graduate was taken on, followed by another, and qualified surveyors were also recruited. The partners achieved the high financial rewards from their vision and commitment. The capital value of the business is also increasing with a view to eventual sale or merger.

Residential and commercial practice — small town: Anna Sharpe

A small town lacked its own chartered commercial property surveyor/practice, reflecting the small amount of commercial property in the area, and the limited level of instructions thought likely. Having had experience many years ago in a residential lettings agency, and more recently in residential development agency with a small private practice, Anna, at 47, considered there to be scope to establish a practice providing both commercial and residential property services. Having moved from the south east, due to her husband's work relocation, detailed research was needed, including an assessment of the success of existing small firms in the area. A small shop was leased at the end of the main high street, ensuring prominence that would help raise awareness, but generally minimising property costs. One key aspect of the development of the business was a link with a small firm of building and quantity surveyors in the area regarding surveys of houses and more detailed aspects of building construction, maintenance, etc, for commercial property. The estate agency work, and the related marketing of services, secured leads for surveys, undertaken by the building surveyors, who in turn provided referrals from their clients in respect of general practice work. The desired amount and type of work materialised, and another benefit of links with other firms has been social contact with other professionals, as well as secretarial cover during holidays.

Retired university lecturer: Janet McMillan

Having taken a redundancy as a lecturer in planning, and holding dual qualification as a member of the Royal Town Planning Institute (MRTPI) and the Royal Institution of Chartered Surveyors (MRICS), Janet, at 59, wished to combine activity in practice with community activities, and new duties as a grandma. Arrangements were established with a local planning consultancy, the local planning department and a local firm of chartered surveyors for Janet to provide technical updates on planning matters in an informal training capacity — in terms of both resource material, and informal discussion sessions/seminars. This also enabled contact to be maintained with old university colleagues, and of particular interest was the involvement in the planning consultancy's case work. As financial rewards were a secondary consideration to Janet, and her costs were negligible, the three clients considered Janet's approach one that they could not refuse.

Young surveyor: Afzal Mohammed

Afzal, at 27, had been qualified for four years and gained experience in a local authority property team and a small private practice in the city where he had studied. It had been the intention to return to his home city to be closer to family, and to help with the development of the family's small property investment and development business. However, this only provided less than one day's work per week, and apart from key advisory input regarding acquisitions, the work was largely administrative. Full-time work was not feasible in view of the outside interests, and self-employment was pursued. Registrations with recruitment agencies for part-time work and contract work, and notifying local councils of the availability for project work, maternity cover, etc, led to sufficient work being offered. This was supplemented by a small client base generated initially through family and friends. In view of the number of clients, the operation was self-employment rather than a salaried capacity. After a year of self-employment, Afzal is continuing to gain a balance of further experience with a view to relinquishing recruitment agency and council work within the next two to three years, and setting up offices which will help the business expand its client base.

A more detailed example of the start to a new business is included in Chapter 8, and Chapter 11, Accounting, provides further illustrations by way of examples of financial information.

Personal Qualities and Working Methods

Having concentrated in Chapter 1 on the opportunities for setting up a new business, this chapter examines personal qualities and working methods in more detail — including their effect on profitability.

Self-motivation

Self-motivation is a key strength in a self-employed surveyor. Surveyors considering self-employment should be able to judge whether they possess the necessary characteristics, and should also consider whether they have the personal qualities including resilience, patience, mental strength, stamina and independence (as well as factors such as technical proficiency, an ability to win business, sufficient business and managerial acumen, and a suitable financial position).

Greedy dreamers do not get far in business, neither do pseudo-entrepreneurs, both of whom seem to have lots of ideas and stories of other people's ventures, but not the talent or tenacity to put things into practice themselves. The motivation to become self-employed should not solely be the potential financial returns. However, if other requirements are met, a mindset of self-employment being a route to riches provides a powerful psychological driver for the venture.

Personalities and patience

Personalities who give up easily are unlikely to be suited to self-employment. Patience is needed, maintaining belief in the vision that led to the business being created. It takes time for instructions to be won, and there will be a multitude of frustrations and disappointments which require a positive response. Lessons will have to be learnt, and rejections from potential clients cannot be taken personally. It is just part of self-employment that prospective clients' representatives will usually be in settled jobs, unaware of the self-employed surveyor's possible anxiety as to whether an instruction will be won, when they will be informed either way, what disruption and difficulties are caused while waiting, and having to juggle other possible future commitments.

A self-employed surveyor does not really need an outgoing personality, but does need to be able to get on well with people, especially clients, and also anyone with whom business is conducted. For most work undertaken by self-employed surveyors, clients seek professional knowledge, honesty, integrity, loyalty, and discretion. However, in certain markets — West End agency for example — the right style, demeanour and social contacts, and the ability to close deals, is key. In the same way that conventional recruitment includes managers taking on people in their own image, and with whom they have something in common, clients sometimes do likewise, and the best surveyor does not always get the work.

Sources of motivation

The more varied the work available, the greater the motivation is likely to be. A greater proportion of working hours are enjoyable, and overall satisfaction and financial rewards are higher. It is therefore worthwhile when establishing the services to be provided, geographical area to be covered, etc, to consider factors such as the variety of work achievable, which types of work involve contact with other people through meetings and site inspections, which type of work involves long hours confined to the home/office, and what line of business enables involvement to be taken in local RICS and any other activities.

Weeks and days should be structured to provide variety. One week could contain numerous meetings and site visits, while another reserves time for reports, business development and possibly leisure activity. A 70–80 hour week could be followed by one of 30–40 hours. All such factors represent the difference between busy and diverse 10–12 hour days, and days characterised by a struggle to get started and an eagerness to stop early. There will be always trade-offs between undertaking the most profitable work, and undertaking lower value work, but which adds to variety and motivation — and perhaps also still leads to optimum profitability overall.

Time management and planning

The above comments are an example of how self-employed surveyors find their own working ways. Consideration also has to be given to time-management, personal management and motivation in order to get the best out of the business, and gain personal satisfaction.

Time-management theory suggests that work which is of high importance and priority should be done first. However, some self-employed surveyors prefer to begin the day clearing non-urgent, straightforward items. This helps to get started, particularly if beginning early in the morning, and psychologically creates a focus for the urgent work. Also, if the priority work was undertaken first, the surveyor would not work until late afternoon/early evening on the non-urgent work, but if the non-urgent work was done first, the momentum with the priority work would be maintained, deadlines met, and longer hours and higher income secured. There are, however, occasions where priority work has such pressurised deadlines not to risk undertaking non-urgent work instead. Even then, some self-employed surveyors consider that they perform best, and more productively, when work is urgent. Adrenalin should be inspiring, and give a power to perform. However, if work is left too late, there is not time for the surveyor to reflect on the issues — whereas if progress was made earlier, thinking time would enable any further ideas and opportunities to be incorporated.

Higher productivity is usually achieved when the mind is fresh, and tiredness results in tasks taking longer. For a self-employed surveyor wishing to undertake only 20–30 hours per week, the above factors are not, of course, as relevant as for the self-employed surveyor who is very determined to build a business, and is prepared to commit to long hours in the early stages.

If working from home, it is preferable to remain reasonably confined to the designated office during the day, rather than using the kitchen, lounge, dining room, etc, including watching television while working. If using all rooms, it is likely to be more difficult to mentally break from work at the end of the day. Sole traders report shutting off the working day by closing the office door, and ensuring all work is out of site.

The examples above are just a few illustrations of how the self-employed surveyor deals with potential distractions, and ensures that work is undertaken as productively as possible, and motivation is generally maintained. Surveyors in salaried employment generally need to see out their working times, and possibly be pressured into obligatory overtime, but the self-employed surveyor has the easier temptation of ending the day early. However, in busy periods, and when particular clients have urgent requirements, the self-employed surveyor is more likely to be inclined to work through the night in order to deliver on time. Sometimes urgent requirements coincide, and working hours become excessive.

Necessity of long working hours

It is often commented that self-employment involves long working hours, especially in the early stages. However, many self-employed people put in long hours even when there is not an extensive amount of work to do, the work is not particularly urgent and they have a satisfactory financial position to avoid long hours. Sometimes there are genuine emergencies, high volumes of work and urgent work, but often the long hours are due to psychological factors. Surveyors setting up a business on a full-time basis are simply keen for it to succeed, and wish to do the venture justice. As with surveyors in regular salaried positions who consistently work excessive hours on the pretext of urgency, the actual lack of necessity is highlighted by the hours regularly averaging the same amount per week.

In the early stages of a self-employed venture, it is possible that work becomes all-consuming, and even obsessive. Days off, and even weekend breaks and holidays in the early stages lead to frustration, and working hours therefore become more enjoyable than leisure hours, especially if the business succeeding and there are exhilarating early moments on which to thrive. Care needs to be taken that such a pattern does not continue beyond the early stages of the business, or at least a targeted period of time. If things go well, there is a danger of becoming a workaholic. Work should only be everything for so long. Life needs balance, otherwise when the work stops and the money is in the bank, there is not much else. People who work excessive hours sometimes report that whereas they were once good contributors at social occasions, they now have little to offer because of their relative isolation from non-work activity, and as all their thoughts are confined to work.

Focus and determination are key qualities in the early years, but as the business becomes established, the self-employed surveyor gains increased confidence as to the certainty of work always being available, and is able to ease off. Complacency still has to be avoided, an eye maintained on possible threats to the business, and clients served with the same tenacity, but at the same time, a different lifestyle is able to emerge.

Stress and exhaustion

Self-employed surveyors who take the development of the business to excess, either because of the volume of work, or because of the intense working methods preferred, need to monitor their health, and generally ensure that they sleep well and eat well, and take suitably timed breaks from work.

External activities are important in order to think about things other than work. There will be periods of high stress, but as the self-employed surveyor has relative control over work and working methods, compared, for example, to a salaried employee, stress should be more easily managed. Financial and family pressures may add to difficulties, in which case, stress management techniques are important, and possibly also medical advice. Excessive hours are likely to lead to exhaustion, but can be managed under limited periods. Longer term exhaustion could lead to mental exhaustion/nervous exhaustion which creates an agitated/nervous appearance with clients and others. An occasional alcoholic drink aids relaxation, but too much slows down the mind, and acts as a depressant.

Professional and mental performance — and "thinking time"

In a similar way that physical performance is enhanced by suitable physical training, an underestimated aspect of an individual's personal development is the ability to enhance professional performance through improved mental sharpness. Facets of this include becoming more incisive in decision making, undertaking work more quickly, being able to deal with complicated issues, having mental stamina in terms of the daily hours committed and being highly articulate both verbally and in print. This is achieved naturally through a high volume of work and challenging subject matter, and by design through working at high tempo.

People who are successful in business tend to see the big picture, and instinctively understand how all the component parts of a venture combine in a strategic sense. Lateral thought processes are required. Thinking time is a key element in developing a successful business. Here, thought needs to be given to the achievements to date, areas for development, opportunities to improve certain areas, and any mistakes being learnt from. When busily dealing with day-to-day work, new thoughts for the business are less likely to be inspired than when the mind is allowed to either concentrate on strategic thoughts, or just allowed to wander in a relaxed environment. Opportunities for thinking time need to be pro-actively created if not arising naturally.

When people are not pre-occupied by particular tasks, they think about things: sometimes intensely, but on other occasions in a light and vague way. Health or financial worries, or business troubles are typically the immediate thoughts, as are leisure activities of sufficient impact. If too many such thoughts come ahead of thoughts about the business, then key thinking time which will be good for the business is lost. People who are generally content with their own company are more likely be deep thinkers about issues, and consequently are well-positioned to develop their own business. In contrast, people who are pre-occupied with their social commitments deny themselves good quality thinking time that will enhance their business. On the other hand, the opportunity to discuss issues with others inspires many ideas which would not have been possible in one's own company, and it is necessary to achieve a suitable balance.

As indicated in Chapter 1, another element of self-employment is the relative isolation compared with a salaried position involving travelling to work everyday and being with other people. People who enjoy being with other people and do not enjoy their own company for very long are not suited to some forms of self-employment. The self-employed surveyor soon also finds that the extended discussions at meetings, coffee breaks and other "down-time" enjoyed in salaried employment now represents lost fee-earning time. Self-employment also means making decisions without the opportunity to discuss the issues through with a colleague.

FOCUS

As indicated above, focus is an important part of establishing a new business. Surveyors who wish to be particularly committed to a new venture, and put in long hours in the early stages, need various qualities, including focus, motivation, energy and stamina. When there is a choice between work or leisure time, and it is tempting to ease off, in the majority of cases the decision needs to be made to work. Such an approach cannot be maintained on a long-term basis, but in setting a target of one, three or five years, for example, the business is afforded the major commitment that its owner wishes to devote.

Self-employed surveyors should set themselves targets and milestones, such as a particular level of profitability, the stage when the mortgage is paid off, when other stages of wealth are attained, or undertaking so many years of hard work before easing off. Such lights at the end of the tunnel provide a longer term focus which supports short term motivation — just the same as looking forward to an imminent holiday and being able to work with enhanced commitment in the meantime.

Entrepreneurial mindset

Academic studies, and also work in a systems based environment, stifle entrepreneurial and instinctive thought processes. It takes experience and insight into the real/commercial world for thoughts to break free from mechanised processes. Surveyors looking to establish their own business should read appropriate business literature and generally become attuned to business, rather than looking to run a business in the automated way that possibly characterised their previous experience.

Being alert to downturn

Even when the workload is high, and the venture is seemingly successful, the continuing availability of work cannot be taken for granted. Similarly, lower value and/or less interesting work, perhaps for clients secured in the early days, cannot be afforded lack of importance and priority. The fortunes of a small business can change quickly, and instructions dry up. If a client has to re-tender work, further instructions might unfortunately go elsewhere, despite the excellent track record of the surveyor/small business. Clients also benchmark their consultants, and provide further work to those considered to be performing best. Complacency can be costly. Clients' business plans also change, and property requirements become less extensive. It is important for surveyors to account for worst case scenario in their ongoing review of their business and its future plans.

It is also important never to forget the humble beginnings of the business, and that the opportunities provided by the smaller clients many years ago, created the track record which led to the higher value work of today. It is simply a healthy outlook which staves off "attitude" which could be detrimental. Compare, for example, highly successful people whose attitude ensures they are widely disliked, to their equally successful, but unassuming and popular contemporaries. It is a small world, and the person left disgruntled from dealings last year, could be in an influential position this year.

Finance and Cash Flow

For a surveyor considering self-employment, business finances are interrelated with personal and family finances.

As mentioned previously, a small professional services consultancy does not need substantial start up capital (ie in contrast with the investment capital and working capital needed in many retailing and manufacturing businesses). Provided the business concept is right, and the surveyor is alert to the range of issues requiring attention, financial obstacles should be relatively limited.

Variations in the cost base

As an illustration of the impossibility of answering the question "how much does it cost to start up?", and as indicated in the preface, this could range between £1,000 or less to £50,000 or more. Contrast, for example, two situations:

- A senior surveyor leaving salaried employment, and handing back the company car, needs to acquire a car which provides the right profile at, say, £25,000, rent offices and undertake fit out at £10,000, buy furniture at £3,000 and a computer, camera, and other IT equipment, etc, at £5,000 — with other costs taking the total to around £50,000. From the examples provided in Chapter 1, this would be similar to Anthony Gallagher's situation, p 18 (and also to the example in Chapter 8 — noting also the cash flow issues associated with income and expenditure, as well as actual capital requirements).
- A surveyor (such as Afzal Mohammed p 21) has the necessary vehicle and equipment to begin with, intends to work from home, and needs only essentials such as stationery and professional indemnity insurance.

Marketing costs are avoided for some new businesses, but for others could be in the region of £5,000–10,000 or even higher. Similarly, staff recruitment costs are not needed for one business, whereas another appoints a recruitment agency at at £2,000–3,000 to provide general advice, and also find a secretary and a qualified surveyor.

If purchasing a commercial property, rather than taking on leased premises, sufficient equity is required (as even when requiring a mortgage, the bank will only be prepared to lend against, say, 75% of the value — ie the loan to value ratio). A property purchase is unusual though for a sole trader, unless perhaps as part of personal investment plans and/or personal financial planning.

The following financial issues are examined below in order to provide a general overview: cash holdings and other assets, business cash flow, domestic mortgaging, personal financial planning, raising business finance and utilising sources of information. A further example of start-up is shown in Chapter 8, to include more detailed personal financial circumstances, and issues to consider if raising finance.

Cash holdings and other assets

When Anthony Gallagher established his business (p 18), he enjoyed a strong cash position reflecting the accumulation of year-on-year savings. His children were all working, and his mortgage paid off, with weekly overheads confined to general living expenses and social activities. Finance did not need to be raised, and cash flow was not problematic, although taxation planning and personal financial planning were necessary. As shown in other chapters, a strong financial position provides flexibility in respect of tax planning, although this depends on the level of profits achievable for the business. This includes consideration of whether sole trader/partnership or limited company status is preferable (see Chapter 5).

Redundancy payments or family inheritance also account for finance not having to be raised. For surveyors setting up business in recent years, the rapid increase in house prices (as well as equity built up over the years) has enabled re-mortgaging to comfortably provide the necessary finance, without the need for business finance.

In contrast to the above situations, some of Michael's contemporaries were fully mortgaged and had other debts (largely reflecting status-conscious social lives and spending habits). Although finance could still be raised, the lifestyle, mortgage commitments and weekly overheads which were feasible with the salary enjoyed in regular employment, would need to be trimmed if a new venture was to take place, and the sacrifices were not considered worthwhile.

Business cash flow

Although the initial capital outlay need not be substantial to set up as a sole trader, certain lines of work take time before fees are be billed (such as a drawn out rent review or rating appeal), whereas others see instructions completed in a matter of weeks, and invoices issued immediately (such as valuation). There is also time between invoicing and receipt of payment. Property management work, and the collection of rents and service charges, should provide regular income through the authorised deduction of fees. When property, secretarial and other costs are incurred, there is a two-fold effect of expenditure taking place immediately, while income is not received for some time.

Self-employed surveyors do not tend to suffer significantly from clients failing to pay fees due (ie bad debts). Although, on one hand, small surveying businesses attract smaller clients which are more likely to default, the more personal working relationships typically established between the surveyor and local business minimise the chance of default. Fees, in whole or part, could be secured in advance if necessary.

Some clients may be may be slow in settling invoices, and the sole trader may have to spend time chasing payments — especially if there are a large number of clients providing small amounts of work, and the surveyor needs the income. Only so much pressure can be applied to clients without jeopardising future work. Even if legal action commences, it takes time and incurs cost, both in terms of legal fees and lost fee-earning time (as well as the frustration and distraction caused — factors which accumulate, and diminish the surveyor's motivation and commitment to the business).

The surveyor should be able to judge whether any financial difficulties relate to cash flow and the timing of billing/receipt of payment in a profitable venture, or whether there are more serious issues as to the viability of the business. Businesses often go bust because of a lack of cash/cash flow, as opposed to simply being unprofitable. While a new business takes time to develop, principals need to be alert to the prospect of it struggling to become financially worthwhile.

Other assets can be used as a last resort if cash flow problems are experienced. Shares can be sold relatively easily, but at the risk that the sale takes place in poor market conditions and/or at low share prices. The car used for business usually needs to create a certain profile to clients, but again could be sold in favour of a cheaper vehicle. Finance could be raised on vehicles instead of outright purchase, and use of leasing and/or credit arrangements similarly help. In the case of capital items such as photocopiers, use could be made of local facilities with a view to purchasing the business' own equipment once trading is established. Businesses sometimes struggle to obtain credit in their early stages, but a surveyor in a personal capacity should have fewer difficulties. It is also important to plan for tax liabilities, including income tax, national insurance and VAT.

Domestic mortgaging

As well as the scope to extend a mortgage and/or use the home as security for other lending, re-mortgaging is another option — either with the current lender or a new lender. Discounted rates, fixed rates, interest only facilities, cash-backs, etc, are usually available, and a higher amount of finance could be raised. A flexible/current account mortgage could be beneficial, enabling the mortgage to operate in a similar way to an overdraft, and it may even be possible to negotiate a mortgage break.

Although re-mortgaging could enable a comfortable level of funds to become available for the business, interest could be incurred unnecessarily (although mitigated by holding surplus funds in a suitable deposit account earning competitive rates of interest). Surplus funds (and indeed any profits of the business) could also be used to pay off a mortgage. There could however be penalties — and it should also be checked that the payment reduces the outstanding amount and therefore further interest payments immediately, rather than being held on account until the year end.

Another aspect of being self-employed is that mortgage lenders generally impose restrictions and conditions. It is usually preferable to arrange mortgaging in advance, while in salaried employment. If a partner is in work while the surveyor establishes the business, this guarantees a source of income, and should also help with mortgaging. (Instead of "partner", the term husband and wife is used subsequently so as not to confuse with partnership, and because of taxation and accounting issues relating to husband and wife). Ownership of investment property usually also gives flexibility regarding the raising of finance, as well as providing an income.

Although a family home is available as security for a loan, it could be at risk if things do not go well. Family members are an alternative source of finance, although this is not always considered an appropriate step to take, as business difficulties could lead to family tensions.

Personal financial planning

Personal financial planning is not specific to surveyors starting their own business, and in any event is determined by surveyors' individual circumstances and warrants independent financial advice. As a checklist, the self-employed surveyor would review aspects such as whether the most competitive terms are secured from domestic mortgaging, whether savings are earning the highest possible rate of interest, the terms of a will, inheritance tax planning, capital gains tax issues (such as if investment or shares are held), whether shares and other investments should be retained or sold, whether life assurance is necessary, whether accident/critical illness cover is needed, and whether private health plans are worthwhile.

Words of wisdom commonly stated in the financial press regarding personal financial planning and the various forms of investment and individual products available, include "invest only in products which are understood", "the value of investments can go down as well as up" and "the capital committed could be lost in its entirely". It is also important to recognise that the majority of investment products in some way usually seek to hoodwink consumers. Endowment mortgages and pensions advice from previous years have more recently been shown to be flawed, as has the mis-selling of other products — and there is no reason to believe that products promoted today will not be the subject of controversy and poor performance in years to come. Even for straightforward bank deposit accounts, interest rates are often designed to merit appearance in the tables of top interest rates included in quality newspapers and other outlets, only for the high initial/ introductory rates to soon fall once the bank has attracted sufficient funds. Comment is rarely seen about the merits of cash holdings/deposit accounts as part of personal financial planning, and equity investments are often peddled on the back of historical performance, and a notion that in the long term they provide the greatest returns. As an illustration, the FTSE 100 index reached 6900 in 1999 but in 2004 is currently around 4500 — giving a reduction in the value of investments of 35%, and needing to increase by over 50% in order to break even (although dividend/income returns of around 10% would reduce losses). In contrast, cash held in deposit accounts would have increased in value by around 20%. Taxation influences comparisons, but this shows the risk of equity based investments — especially to a sole trader whose business runs into trading and/or cash flow difficulties, and investments have to be cashed in as the only means of financial support.

People also need to be able to sleep comfortably at night regarding their personal finances, and not, for example, be concerned about the performance of the equity markets. Surveyors should take advice from an accountant and/or financial adviser, but at the same time being aware that independent financial advisers are still in business to make money, and might sway towards advice which yields commissions. As mentioned in Chapter 5, tax planning opportunities, including between husband and wife, also need to be regularly evaluated.

Raising business finance

For the surveyors who cannot bring about a sufficient financial position through their domestic circumstances, business finance is required. For the majority of surveyors becoming self-employed, this involves a loan and/or overdraft from a high street bank (as opposed, for example, to enterprise grants, support from business angels and venture capitalists, investment from family and other shareholders/investors).

Business plans

A business plan is required, and must contain more detailed justification than the entrepreneurial instincts and appetite for hard work which might actually prove to be the most compelling reasons for the venture. As outlined in Chapter 4, Business Plans, the surveyor is looking to sell the concept to the bank, and hoping to secure a satisfactory financing facility at a competitive rate of interest. To the bank, the surveyor represents a new source of business, and as well as business banking facilities, they are able to sell insurance, pension and other services (although they cannot make such requirements a condition of the loan).

As the bank will ask the surveyor/borrower questions based on the business plan, surveyors need to be well-prepared for meetings, and be able to talk fluently through all issues (ie without reference to the business plan or other notes, except for the detail in financial projections). In connection with the business plan and the raising of finance, it could be necessary to give presentations.

With any small business, business experience and a good profile and track record of the individuals concerned is important. Compared with the many prospective new businesses based largely on ideas, rather than expertise, and without a track record, a chartered surveyor wishing to set up business tends to have a good background and well-thought out plan for the business, and represents lower risk (and welcome business) to the bank. The strength of the surveyor's research for the venture, and comments on the markets for the services, economic and property market conditions, etc, in the business plan is important.

Banks' other requirements

A chartered surveyor looking to raise finance almost always has their own cash or other assets/equity to put into the business, or alternatively offer as security. For a start-up business, this is usually critical in securing finance — although it does not mean that the security has to cover the whole loan. The bank generally takes into consideration the other assets, financial standing, wealth, and lifestyle, of the borrower. If, however, surveyors have a poor credit history, such as county court judgments, cheques frequently bouncing, etc, the bank may not be interested in lending money.

The relatively low level of financial requirements for new surveying businesses means that surveyors deal with the bank's small business department. Lending decisions work partly on the basis of box-ticking, and the bank having adequate justification on file — hence the importance of the business plan and financial projections. Although the impression from the bank's business literature is that guidance and support is available from the bank, this is likely, in practice, to be limited to basic immediate banking and financial matters, rather than detailed consideration of the viability of the business plan and business planning. While the bank's personnel work closely with smaller businesses not run by professionals such as surveyors, a surveying business is typically well-placed to be left to itself.

Surveyors should shop around for the best deal, although it is likely to be easier to demonstrate a track record in respect of financial discipline when dealing with their current bank, and this should help secure more competitive terms. Even if the surveyor is intent on working with the current bank, discussions with other banks provide comparative terms which should assist with negotiations. Another influence on lending is the recent experience of the local bank and/or its representative with similar businesses, and if this happens to be bad, then it could be difficult to secure the most favourable terms.

Cost of finance

Business finance is generally more expensive than the level of interest repayable on a domestic mortgage, although unlike purely domestic mortgage arrangements, interest is a business expense which is set against income/profits, and therefore reduces tax. Business borrowing is almost always at higher cost than the amount of interest earned in a regular deposit account. As a broad rule of thumb, loan interest is around 2% to 5% above the Bank of England base rate, whereas domestic mortgages are around 0.5 to 2% above the base rate, depending on the provider and particular products/incentives being promoted (ie 0.5% is highly unlikely to be sustainable).

The rate secured by a new surveying business for a loan depends on many factors, but in broad terms it is likely to be in the region of 4% to 4.5% above base. A particularly low risk venture, offering assets as security, and valuable future business to the bank, could achieve 3.5%, or negotiate well and better this. An arrangement fee is usually also payable in the region of 1% to 1.5% of the loan. The interest rate for an overdraft secured by assets of the borrower would be the same as the rate on the loan, or slightly higher. For unsecured overdrafts, the interest rate could be up to 5% to 7% above the base rate.

If the base rate varies by several percentage points, the margins referred to above would change. It is generally possible to repay loans early, although could be subject to a repayment penalty, such as 1% of the amount repaid. Depending on interest rate movements, the repayment of fixed rate borrowing could be expensive.

Banks usually require loan and interest repayments to begin straightaway, either on a monthly or quarterly basis, although it may be possible to negotiate a capital holiday/interest only basis for a period. Immediate loan repayments as well as interest repayments will, of course, have to be reflected in cash flow planning.

Working relationships with the bank

More important than relatively small savings in interest, is ensuring that the necessary overall arrangements are in place which help the business through its early stages of problematic cash flow. Good working relationships with the bank and its representatives are important — including a supportive rather than profiteering attitude from the bank. In the case example in Chapter 8, good relationships and the establishing of a track record for the business are vital to raising further finance, and minimising the interest rate payable.

Preserving flexibility

It is important to secure a financing facility for more funds than are initially thought to be necessary, and draw down further funds if needed (and as always, the worst case scenario should be contemplated). A loan involves meeting repayments and interest payments, and while providing a guard against a poorer cash position than anticipated, could become costly through unnecessary interest repayments. In contrast, an overdraft facility incurs interest costs in proportion to its need.

The surveyor is in a better position in seeking finance, and competitive rates of interest and other terms, when certainty and low risk is presented to the lender, than when the business has run into problems, and it needs funds urgently. Financial difficulties could also raise questions to the bank as to the competence and credibility of the surveyor, notwithstanding the seemingly high quality business plan which was initially presented. It would be much better to be able to update the bank on how the business has exceeded the initial projections on which the finance requirements were based. This would be particularly beneficial if the business had required initial start up capital with a view to expanding the business by taking on staff, moving to larger offices, opening additional offices, acquiring freehold premises, developing investment interests, etc, at a later stage. Whereas the surveyor's personal finances may have largely supported the initial stages of the business, greater reliance on external/bank finance is now likely to be required. The surveyor should remain conscious of how credibility is conveyed to lenders, and financial planning is important. If the business is going well, it should be possible to negotiate a reduction in the interest rate charged, reflecting the lower risk to the bank.

Flexibility is also preserved by having more than one business bank account, and perhaps having several domestic bank accounts. Although it is important to avoid being over-reliant on borrowing and not being in a precarious cash flow position, the scope to run into overdraft at small levels on a number of accounts should not alert a lender in the same way that a larger overdraft on a single account placed with the lender is likely to. Credit cards could also guard against temporary/last resort cash flow needs, and personal loans could be secured from the many less well known lenders advertising in mainly tabloid newspapers, and on TV.

Utilising sources of information

A range of information is available from the high street banks in respect of banking arrangements, the raising of finance, and also aspects such as pensions, healthcare, life assurance, insurances, and credit cards.

As well as the brochures in the branches, the websites of the high street banks contain the relevant information. The banks also provide economic and business bulletins. For further details on up to date rates of interest on savings, life assurance, etc, the *Financial Times* website is *www.ft.com/yourmoney*.

Chapter 4 examines the contents of the business plan, and after Chapters 5 to 7 have considered issues in respect of business status, winning business and business growth, examples are provided in Chapter 8 of how business and financial planning interrelates in the case of a new business looking to grow rapidly.

Business Plans

Whether or not a business plan is needed in order to help raise finance or support other requirements, it is helpful for the surveyor to still commit detailed thoughts to paper, and generally refine the business strategy. It is important to consider, in detail, the market for the services, pricing and profitability issues, the preferred working methods and financial/cash flow or other constraints — including risks to the business. This chapter considers the general content of business plans, and Chapter 8 provides a case example, paying particular regard to the need to raise finance.

Importance of detailed planning

New businesses do not always give sufficiently detailed consideration to business planning, often reflecting the enthusiasm to actually begin trading, and put ideas into practice as soon as possible. This increases the likelihood of the business strategy being frequently changed, and generally being wayward — with more problems tending to arise until the business settles into its natural market and its most appropriate management approach. Instincts are important in business, and help an edge to be secured, but this does not mean that instincts are not properly evaluated when strategic planning is undertaken.

In setting up businesses, there is a risk that its principals/entrepreneurs are deluded as to the likely success of the venture. The ease of winning business is easily underestimated. Bad judgment also derives from over confidence, pressures, enthusiasm, exuberance, wishful thinking and naivety. Instincts create opportunities which appear worthwhile, but once more detailed thought is given to how markets might be entered, at what cost, etc, the ideas have less merit. Discussions with others may help constrain any adverse consequences, although arrogance and pride also prevent the necessary advice and counsel being sought. As mentioned previously, it is always important to contemplate the worst case scenario.

Although it is notoriously difficult to effectively plan the success of the business from the outset, and it can meander in unpredictable directions, the business plan helps provide a considered check on viability. This is particularly useful in the case of partnerships where there is potential conflict regarding the activities, markets, etc, that the partners wish to pursue. Another benefit of the business plan is that, once settled, it should help increase confidence in the proposals for the venture.

The simple need to prepare an articulate written plan helps the surveyor/ partners to be more thoughtful and incisive, and the business to be more focused. Time is needed for reflection on ideas (with "thinking time" being important for principals of businesses, as mentioned in Chapter 2). Having colleagues or a bank manager provide views and ask numerous questions helps ideas and plans to better come together, and any weaker aspects to be sifted out. As with established businesses, managers, or an entire business, sometimes miss certain issues and opportunities (ie blind spots). Successful business people see things that most others cannot.

Uses and variations of the business plan

The business plan is helpful for prospective partners, staff and investors, to demonstrate the growth potential of the business and the opportunities it provides. A business plan should be regularly reviewed and updated as part of the ongoing evaluation of the success and direction of the business. Different versions of the business plan may be prepared for lenders, partners or for other reasons in view of the different purpose, level of content and technical detail needed. A composite business plan could work well by the effective use of appendices containing the more detailed technical/surveying information which otherwise clutters proposals to a lender if included in the main body of the report. An alertness should always be maintained to the content of the business plan in terms of the view of lenders. It may be preferable, for example, not to comment on the more remote risks which represent worst case scenario, but which can unnecessarily induce concern to a lender.

Presentational issues

The business plan conveys the calibre and credibility of the surveyor/partners to its intended audience, and professional help may be needed in its formulation, notwithstanding the surveyor's familiarity with report writing. Financial information enhances the image presented (however difficult it might be to make accurate projections) and shows a lender, or others, the depth to which the surveyor is able to analyse and plan the business, including to reflect potential risks.

The business plan needs to be well-presented, and carry the right image. However, an impression of "style above substance" must be avoided, and the latest design and print options are not essential. If business stationery is already available, this adds to the presentation. Font type and size, line spacing and margins need to be suitable. Stationers and printing shops are able to advise on the options available in respect of presentation and binding.

As with all business reports, it is important to "write for the reader". Apart from the basics of not inadvertently assuming that the reader has knowledge and covering points in a suitable sequence (not leaving the reader turning back pages thinking that they have missed something), regard needs to be given to the level of property expertise that the reader has. Surveyors/partners do not need an explanation of property issues, industry jargon, etc, but for a bank, technical commentary should be avoided, and key terms and issues explained in an appendix/glossary if need be. Also, headings in a report/business plan need to be suitably expressive and add to its high quality structure. The business plan should be read by other people, including family and friends, as well as property and other professionals, as this helps ensure that the report is easy to understand, and does not rely on assumed knowledge.

Information that is omitted could appear careless, and aspects of the business plan which are unrealistic/over-stated could cause the reader to question the accuracy of the whole document, and the credibility of the surveyor.

Content of the business plan

There is no standard format for business plans, but there are conventional points and a broad order to be followed. The content of the business plan is driven by the nature of the business, the audience of the business plan, and what the business plan seeks to achieve (such as getting money from a lender — achieved by conscientious and well-articulated proposals which convey the low risk, certain outlook for profitability and the expected success of the business). Banks do not require businesses to be making substantial profits, and be the next great growth story that leads to stock market flotation — but if the business plan is for venture capitalists and/or other investors, potential riches are key drivers.

A cover page could include the name of the surveyor/partnership/business and a title of the business plan. A title page could repeat this and additionally include the name of the surveyor/partnership/business, followed by details such as address, telephone number, fax number, e-mail address, website address, and the date of the business plan. A contents page could then follow. Pages need to be numbered, and a numbering system (1, 1.2, etc) adopted for sections and paragraphs, as users of the business plan often cross-reference, and look to pick out key aspects.

An executive summary should comprise a single page of key points. This is usually the last part to be written, and comprises a concise list which is captivating to the reader, and accurately conveys the essence and perceived potential of the business, together with its alertness to risk factors.

The main part of the business plan begins with an overview on the nature of the business, its market, office location (or home working basis), legal status, the number of people involved, etc.

Profile of principals

Details of the surveyor/partners need to be set out. The background, track record, achievements, qualifications, credibility, calibre, etc, of the individuals are vital. Comment should be made on matters such as the expertise available, clients already secured, how partners complement each other, how responsibilities are to be shared, etc. Further detail is provided by including CVs in the appendices of the business plan.

Market issues and services

A commentary needs to be provided on the market, the services the practice intends to offer, the type of clients sought, competitors, market pricing, advantages and disadvantages against competitors, how services will be differentiated, how niche concepts will be exploited, how personal service is key, how fees will be competitive and margins high because of the low cost base, how business development will be undertaken (and how the surveyor/business considers they are able to break into an established market), the clients already on board, etc. Comment should be made on how price sensitive services are, or, for example, whether the market is likely to bear relatively high fees compared with competitors because of the profile and track record of its principals.

Business development and marketing

Regarding business development, a more detailed commentary could be prepared in respect of marketing, even extending to a separate marketing plan or dedicated appendix. This outlines how the business is to be marketed, and at what cost — and would develop the points made in earlier chapters regarding the type of clients, areas of practice, etc, which are sought. In the early stages of a business, there is common tendency to take on any work that is available, but which does not represent the areas of practice, type of client, level of fee, etc, that will help the business develop. The business/marketing plan helps ensure that only the right business is won, and the longer term direction of the business kept on track.

Further detail on marketing is included in Chapter 6, Winning Business, Chapter 7, Business Growth, and Chapter 8, Case Illustrations.

Opportunities, risks and strengths,

SWOT analysis (strengths, opportunities, weaknesses, threats) is inappropriate as a heading or phrase, as it can appear as if it is lifted directly from a text book, but such factors still need to be highlighted. It is important that the risks of the venture are evaluated, and while brief comment is included in this section of the business plan, the surveyor's alertness to downside factors could also be shown as part of the financial projections provided, the funding requirements identified and a section such as "risks to the business" (although "risks and further opportunities" enable upside potential to also be demonstrated). Risks range from economic and property market fortunes to issues such as the level of work resulting from a rating revaluation, or the effect of possible changes in legislation. Other aspects which could be highlighted in the business plan include the areas of experience/expertise of the surveyor which help open new markets if need be — especially if the business is initially focused on specific areas not representing the full breadth of the surveyor's expertise. Dependence on key clients, and the potential for increased competition also warrant comment. For a larger practice, the ability to attract staff of the right quality, the salary and other terms they demand, the ease of replacing staff, etc, are relevant (see Chapter 7, Business Growth).

Capital and running costs

Details need to be provided on requirements in respect of premises, other significant capital/investment items and staff, together with an overview of the general running costs of the business. Examples of costs were given in Chapter 1, and are also shown in Chapters 8 and 11.

Financial projections

In respect of financial projections, comment needs to be made on the potential level of fees achievable, capital/investment costs and running costs. It is helpful to distinguish between fixed costs (which are the same irrespective of fee income /turnover) and variable costs (which are incurred broadly in line with the level of turnover achieved). Fixed costs need to be exceeded before any profit is made.

Although a difficult exercise to undertake with accuracy, projections are ideally made on the level of fee income that could be achieved on a month by month basis. It is not however, just the receipt of fee income which needs to be shown, but the level of instructions being received and the value of the activities being undertaken on a work in progress basis (noting how the timing of income varies between different surveying activities — as mentioned on p 12). In the early months, the level of instructions and number of chargeable hours are likely to be low, but in due course should increase. Indications can be provided, assuming, for example, a working week of 40 hours which includes 30 chargeable hours, also allowing time for holidays, and other time off. An indication is needed of the approximate rate per hour chargeable. Even fees based on a percentage of rent or sales price should still equate approximately to a rate per hour in order to ensure that such jobs are worthwhile. Optimistic and pessimistic scenarios could be constructed, and statistics produced such as the number of working hours/chargeable hours needed in order break even, and to achieve a particular level of profit. Fixed costs and variable costs are relevant here. The proprietor may have stated aspirations to commit 60 hours per week to the business for one year, three years or five years and this facilitates related income/profit projections. Such personal aspirations should also be stated somewhere in the business plan (see also, medium and longer term issues below).

A projected profit and loss account could be provided (as could a balance sheet, although this would not, for example, be of much relevance if profits were withdrawn, and assets remained largely the same as the start of the year). In addition to the usual basis of accounts which would be suitable for the tax authorities as part of annual accounts, illustrations could be provided on the basis of work actually undertaken. (Work in progress/uncompleted work would not generally have to be included in the sole trader's company accounts — although work instructions which are completed but not invoiced would be. Debtors at the end of the year who have been invoiced would, of course, count as income in the current year's accounts).

Although the projections are, in practice, subject to various tolerances, and other factors, the figures still provide a helpful guide to lenders, and enable the surveyor to create measures of activity and profitability which should serve the business well year on year.

Cash flow

It is important to comment on how particular aspects of the business affect cash flow. This is in line with comments made on p 12 about certain areas of surveying producing income relatively quickly after the initial instruction and completion of the task, whereas other instructions do not yield income for many months. If the business plan shows the business to be profitable, the bank should be reassured as to the low risk of lending, and be relatively relaxed where the need for finance reflects cash flow trends for a particular type of business, rather than trading difficulties. Cash flow projections are therefore needed, mapping out the point in time when expenditure is incurred, and income received.

Any points at which cash flow is likely to be problematic should be highlighted, although the nature of the business is likely to involve reasonably regular income (unlike certain other trades where cash flow experiences seasonal variations). Cash flow variation throughout the year does, however, depend on whether the firm has a small number of large clients who represent a considerable proportion of total income, how frequently invoicing takes place, and how soon payment is made. A large number of smaller clients, and more frequent invoicing, should provide more regular cash flow. Many businesses work to a financial year-end of March or April, and are either keen to make payments, or wish to delay payments, around the year end.

Cash flow projections help the business establish its funding requirements (which could be the next section in the business plan). In addition to the usual items of business income and expenditure, cash flow projections need to include allowance for the surveyor's/partners' drawings from the business (unless this is not necessary for some time because of the state of domestic finances), and also for loan and interest payments. The various figures adopted and projections made are likely to need supporting statements.

Liability for taxation also has to be reflected in cash flow projections. Chapter 11 provides information on the point in time that income tax and corporation tax will be due, and Chapter 10 comments on VAT, including the benefit of cash accounting.

A business plan prepared for purposes other than the raising of bank finance requires less financial information. However it is still useful to include profit projections and analysis of different services in terms of fees, chargeable hours, non-chargeable hours, fixed and variable costs, cash flow and profit/margins.

Funding requirements

Funding requirements need to be summarised, and this section should include comments on the level funds committed by the surveyor/partners. The surveyor's/partners' personal financial circumstances should be set out (as should be any other business interests earlier in the business plan). As mentioned in Chapter 3, it is better to secure a greater finance facility than is likely to be needed. New businesses cannot underestimate the time it takes for the business to develop, and for invoices to reach the point of issue, and then payment. In outlining their personal financial circumstances, surveyors need to show that if the business runs into cash flow difficulties, or even problems regarding the level of instructions won and profitability, funds from other sources can easily be realised (such as re-mortgaging the home, cashing in investments, using other savings, or other family funds being available). Financial issues were covered in Chapter 3, and further detail is provided within the case example and illustrative business plan in Chapter 8.

Medium term and longer term issues

A business plan inevitably focuses on the prospects for the early stages of the business, but it is also important to identify the medium and longer term aspirations. A sole trader could aspire only to work on a self-employed basis, and avoid the issues which arise when taking on other staff. Another surveyor may wish to bring partners into the business at a certain stage, or look to open second and further offices in other locations. This could be particularly attractive to the bank in view of the further lending business that could be available, as well as the increased scale of operation adding to the level of general banking charges. The business plan could also incorporate a strategy relating to timescales — such as a plan for year one, years two to five, and thereafter.

Managerial issues, human resources, business growth, etc

Issues in connection with business growth, including managerial aspects, are covered in Chapter 7. The business plan could provide brief comment on the issues anticipated, and the systems likely to be implemented in order to bring about well-managed and successful expansion. This also shows the sole trader's capacity to manage a business — as opposed to simply being a good case surveyor venturing out alone.

A summary/conclusion can also be provided to the business plan.

Dealing with difficulties

If things have not gone well in the early stages of the business, the business plan needs to be reviewed. The concept and direction might not be at fault — rather it may be that longer time is required to develop the business, to turn leads into instructions, and for marketing to attract the right clients, etc. Some sole traders return to regular salaried employment because of financial needs, but others return through impatience that the business is not developing. Sometimes, however, surveyors just have to cut their losses, and accept that, for whatever reason, things have not worked out.

Confidentiality

Another issue regarding the preparation of business plans for banks and others is the confidentiality of the information provided, and that this does not in any way get into the hands of competitors or potential competitors. Confidentiality statements are possible, but might not be to any great practical effect in preventing anyone passing on ideas. Dealings with banks, however, should not create any risks, but in the case of information for prospective partners and other investors, there is a difficult balance to strike between selling the potential, and not opening others' eyes to opportunities which they could pursue alone.

Business Status and Tax Issues

The status as sole trader, partnership or limited company affects matters ranging from the profile of the business, to its financial affairs and the tax efficiencies which are achievable.

The first part of this chapter explains the main advantages and disadvantages between these principal types of venture, and summarises other issues involved when setting up a surveying business. The second part considers income tax, national insurance and corporation tax issues, including the scope to secure tax efficiencies.

Beginning as a sole trader

Most surveyors setting up their own business begin as sole traders. A sole trader is the sole proprietor/sole practitioner/principal of the business, and remains so even if employing other people.

Personal liability

The sole trader is personally liable for any losses incurred by the business, which could extend to the loss of the family home and other assets. In order to guard against the worst case scenario of bankruptcy, the sole trader can ensure that suitable assets are held in the name of a husband or wife.

Informing the Inland Revenue

On commencing self-employment, the surveyor needs to inform the Inland Revenue. At the time of writing, the "Helpline for the Newly Self-Employed" is 08459 15 45 15 and information on setting up in business is available at *www.inlandrevenue. gov.uk/startingup.* If these details change, a local office Inland

Revenue office will provide the relevant contact points, together with details of advice helplines, publications, website references and other guidance. There is no shortage of government and other business literature when setting up in practice, although it is not always easy to find material specific to the circumstances of the business being established.

For national insurance purposes, the above registration with the Inland Revenue needs to take place within three months of starting the business, otherwise a penalty of £100 is payable. Registration involves completing a form (CWF1) with brief personal and business details. This commences billing by the Inland Revenue for class 2 national insurance contributions (£2.05 per week, payable £26.65 per quarter, based on 2004–05 rates). It also ensures that a self-assessment form to complete the year-end Inland Revenue income tax return is sent in due course. Even if surveyors have been making a self-assessment return while in regular salaried employment (such as to account for other income, or investment income, if a higher rate tax payer), registration is still necessary.

Administrative requirements

Status as a sole trader involves relatively little administration. As summarised on p 136, records have to be maintained in accordance with Inland Revenue requirements (and possibly Customs and Excise requirements regarding VAT). Records of income, expenses and capital expenditure are needed in order to calculate profits which are subject to income tax and national insurance, but only the Inland Revenue's self-assessment forms need to be completed.

It does not really matter if the sole trader uses a combination of domestic bank accounts and business accounts for trading. Some sole traders use only domestic accounts, and, in doing so, avoid bank charges. The business name, or at least the name on invoices and clients' cheques, would usually need to be the surveyor's personal name in order for cheques to be banked. However, while some banks do not raise concerns about business payments being made to a domestic account, and might even allow payments to certain trading names to clear the account, a high number of payments could prompt a request to establish a business account.

Although such informal arrangements would suffice in the early months of business when there are so many other matters needing attention, it is preferable to commence a small business on a more organised footing. A business bank account is usually opened with an initial capital sum, and, thereafter, all income and expenditure is detailed on business bank statements, and corresponds with invoices/receipts for fee income and expenses, and the surveyor's own accounting records of such transactions. Although software packages are available, in the early stages of the business, it should be relatively straightforward to work to a basic Word™ file or Excel™ spreadsheet.

As mentioned in Chapter 11, Accounting, good quality accounting records are beneficial for the raising of finance, attracting partners, selling the business, and securing tax efficiencies. Well-organised records also benefit dealings with the Inland Revenue and Customs and Excise. Although a balance sheet, involving the balancing of all items, including reconciliation with bank accounts, is not necessary for a self-employed surveyor's returns to the Inland Revenue, one could still be prepared.

Publicly available information

In operating as a sole trader, it is not possible for anybody to establish the level of profitability of the business (in the same way that all individuals' tax affairs are confidential). If a limited company is established, details such as the identity of shareholders and certain financial information are recorded at Companies House, and are publicly available. However, as shown in Chapter 11, surveyors establishing a small limited company are likely to only have to file abbreviated accounts which limit the financial information provided, and generally make it impossible for others to gauge profitability (although if profitability is high and profits/dividends are not drawn, the increase in assets could give an approximate minimum indication of the scale and success of the business).

Tax efficiencies

All profits from the sole trader's financial year are subject to income tax and national insurance. In contrast, profits could be retained within a limited company, albeit subject to corporation tax, and paid out as salary and/or dividends at suitably tax efficient points in time (unless, of course, the surveyor is not in a financial position to defer remuneration). More detail is provided later in the chapter, and illustrations are also provided in Chapter 11, Accounting. If there are likely to be losses, an accountant's advice needs to be taken on how these could be set against income/other profits or carried forward to the next financial year.

Business names

A business name/trading name, such as Allan G. Hagley Surveying, or just Allan G. Hagley, could be adopted by the sole trader.

A limited company could also be formed simply to secure the name of the limited company, such as Allan G. Hagley Ltd/ Allan G. Hagley Surveying Ltd. The company does not need to trade, and is dormant. There is however an annual charge of £15 when an annual return is made to Companies House. Dormant company accounts have to be filed, although only a simple form is completed. Trade names and logos are protected by trademarks.

Trading names need to meet legal requirements, and reference can be made to the Business Names Act 1985 (noting RICS requirements also — p 112). It would not be possible, for example, to trade as Jones Lang Hagley, DTZ Debenham Hagley, or King Sturge Consulting (known as "passing off") in order to provide an elevated trading status through association with other businesses.

Profile

For an individual surveyor, or small practice, sole trader status is common, and there is not necessarily any added prestige through being a limited company. The true identity of the business needs to be stated on letterheads, which could be as simple as including "Principal: Allan G. Hagley" at the bottom of the letterhead. Even if a limited company operates, trading names can be adopted — although, again, the letterhead would need to make this clear, for example, that "Allan G. Hagley Surveying is a trading name of Hagley Ltd". In order to minimise the possibility of someone undertaking a search against Companies House records, based only on the surveyor's name, a relatively non-descript limited company name could be adopted. Searches are, however, possible against individual directors — which reveal other companies for which they are a director, and also their personal address — unless it has been excluded for some exceptional circumstance.

Limited companies

If setting up as a limited company, statutory compliance, and accounting and taxation requirements are more extensive, and also incur more cost compared with being a sole trader. This is a private limited company (as opposed to public limited company).

Company formation

An accountant helps set up a limited company, although it is relatively straightforward to instruct a company formation agent do this instead. It is necessary to have a memorandum of association and articles of association, although these are standardised documents which the accountant or company formation agent easily deals with. A new company is usually established, although an "off-the peg"/"off the shelf company" could be bought if necessary, and the name changed. There are unlikely to be reasons for a surveyor to wish do this (and this is less common nowadays as a new company is established in a matter of days rather than weeks).

Director and secretary

A private limited company (such as "Hagley Ltd"), needs a director and a secretary. This could be the surveyor, as director, and husband/wife/partner (or accountant) as secretary. Directors and secretaries of limited companies have certain legal duties. Details are available from Companies House, together with a range of other information (see Chapter 13, Further Information).

Registered office

The company also needs a registered office. This is usually the address of the business, but could be elsewhere, such as the address of the accountant. The tax office of the limited company is in the area of its registered office. As stated in Chapter 9, there are certain RICS requirements in respect of the memorandum of association and articles of association or equivalent constitutional documentation (and which would also apply to "limited liability partnerships" — see p 50).

Share capital, and company stature

A limited company can be established with, say, £100 share capital. Limited companies are known for not necessarily having any stature — although setting up a limited company, and holding positions such as managing director, chief executive, and chairman, may appear impressive. A limited company also needs a bank account in the limited company name.

Personal liability

A limited company limits the surveyor's/shareholder's liability to the amount of capital put into the business, but directors/surveyors could still be sued personally for negligence, for example, or be held criminally liable for the company's acts. Directors may also have given personal guarantees (which are usual with a small company), so there is still personal liability. Directors are responsible for knowing what is going on in the business.

Company stationery

The company letterhead needs to show the registered name of the business in full, either the names of all the directors or none of the directors, the registered address, the trading address and the place of registration (England/Wales or Scotland).

Other compliance

Annual general meetings (AGMs) might also have to be held, although for the small limited company established by a single surveyor principally for tax benefits, such compliance matters are handled relatively easily. It is possible to disincorporate and go back from a limited company to a basis of sole trader, but this is complex, and also involves capital gains tax issues.

Queries

Factual information rather than specific business advice is available from Companies House. A surveyor should not experience difficulties with compliance, but must ensure that forms are competed and returned, as well as accounts being filed on time (otherwise there could be fines and prosecutions).

Partnerships

As with the sole trader mentioned above, the Inland Revenue must be notified in the case of a partnership (by each of the partners individually). The profits of the partnership, in accordance with arrangements regarding profit share, means that surveyors are liable to income tax in the usual way. Retaining profits in the business does not therefore alter the tax liability. As well as the self-assessment tax return mentioned above for sole traders, a partnership/partners have to file a tax return for the partnership. Again, accurately maintained accounts help with tax affairs, and when looking to take on a new partner or perhaps sell the business. In contrast with the limited company, it is not possible for others to obtain information on the partnership's affairs.

National insurance contributions are less for partnerships than in the case of limited companies where remuneration is taken as salary. The terms "partnership" and "partner" refer to the owners or "equity partners" of the business (noting that the term "partner" is also used in some of the larger surveying practices for surveyors who are employees/managers).

A formal partnership agreement does not have to be established, but it is beneficial. As well as details such as the commencement date of the partnership and the nature of the business, issues should be covered such as the basis of profit distribution (including how it is shared among partners and the amounts distributed — noting also that there might be losses), arrangements in respect of partners being added to the business or leaving the business, how much capital each partner commits and the position in the event that partners leave, provisions for resolving disputes, the extent to which partners can be involved in professional activities unrelated to the partnership, and arrangements regarding holidays/sickness.

The partners are jointly and severally liable for any debts of the partnership, including those which relate to the actions of one particular partner.

A limited liability partnership limits the liability of individual partners, and has the taxation arrangements of a sole trader and partnership. An incorporation document has to be completed, and the name registered with Companies House, and an annual return filed. As with a partnership, an LLP does not have shareholders or directors — and has "members" not partners. Detailed advice is available from accountants, although a small firm of surveyors would usually be a conventional partnership.

VAT registration

VAT registration (if annual turnover is greater than £58,000 for 2004–05 — see Chapter 10) is unaffected by whether the surveyor is a sole trader or limited company. It would not be possible to operate both forms of business in order to share income and avoid VAT, but if surveyors have a number of business interests, accountants' advice should be taken on the criteria by which they would be judged to be connected. With a partnership, VAT relates to the total income/turnover of the partnership, rather than to the income/turnover earned by each partner, and which may be below the level at which VAT registration is compulsory.

Taxation of income/profits

The initial areas of taxation that a sole trader needs to understand when setting up a new business are income tax and national insurance. It is helpful to be aware of corporation tax and any benefits that derive from limited company status, even if a limited company is not established from the outset.

Constant alertness should be maintained to general tax planning opportunities, and personal financial planning is important. Tax issues and company status are also relevant in terms of the cost of employing other staff.

This section provides a basic overview of taxation liability, and Chapter 11 provides illustrations in respect of accounting. Figures for the 2003–04 as well as the 2004–05 tax year are provided in order to demonstrate tax liability for the most recent full year.

Income tax

When working in salaried employment as an employee, receiving payment on a four weekly basis, for example, income tax and national insurance is deducted at source by the employer as part of PAYE (Pay as You Earn).

The amount of income tax and national insurance which is deducted depends on the salary of the employee, and any allowances and reliefs relevant for the tax year. Tax years run from 6 April to 5 April, and income tax and national insurance is collected by the Inland Revenue.

For the 2003–04 tax year, the tax payer had a personal allowance of £4,615. This means that the first £4,615 of income is not subject to income tax (ie a rate of 0%). The next £1,960 is taxed at 10% (a low rate introduced to help low income earners). The next £28,540 is taxed at 22% — known as the "basic rate" of income tax. Income above £35,115 (£4,615 + £1,960 + £28,540) is taxed at 40% — known as the "higher rate" of income tax. There are also other aspects to consider such as an age allowance, married couples allowances if over 65, or any other allowances.

The rates for the 2004/05 tax year are £4,745, £2,020 and £29,380 (with the basic 22% rate of income tax beginning at £6,765 and the higher 40% rate beginning at £36,145, assuming no other reliefs/allowances). The bands are increased because of inflation (ie higher consumer prices and higher wage levels). It is politically difficult for the government to raise tax revenue by increasing the basic and higher rates of income tax, so more subtle ways are preferred — often referred to as "stealth" taxes. This could include not raising the personal allowance and other tax bands, in order that, because of annual increases in wages, more of the nation's total income is subject to tax, including at higher rates for some income.

If an employee earned £25,000 during the 2003–04 year, the tax liability would be £4,249.50, calculated as follows:

	£	£
0%	4,615	–
10%	1,960	196
22%	18,425	4,053.50
	25,000	4,249.50

If an employee earned £40,000 during the 2004–05 year, the tax liability would be £8,207.60 calculated as follows:

	£	£
0%	4,745	–
10%	£2,020	202
22%	£29,380	6,463.60
40%	£3,855	1,542
	£40,000	8,207.60

Such a calculation could be undertaken by employees at the end of the tax year to confirm that the correct amount of tax has been deducted. The above calculations should correspond with the information on the P60 form provided by the employer. The final pay slip should also state the total amount of tax deducted for the year. Although deductions are made on a four weekly basis (or weekly, or calendar monthly), income tax works on an annual basis, and adjustments may be needed during the tax year and/or at the end of the year to ensure that the tax liability is accurate. Adjustments may be necessary because of salary increases, overtime, and bonuses. Other benefits, such as company cars and health schemes, are taxable. Pension contributions could also influence the calculations as relief against income tax is available (although calculations should change only for higher rate tax payers).

Income from salaried employment could be relevant during the year in which a surveyor established a new business, and would be incorporated into the end of year self-assessment returns made to the Inland Revenue. Salaried employment could also be relevant if self-employed surveyors held a directorship with a limited company, or under any other circumstances where self-employment included a salary, even if it is for part-time employment. Other income would also be relevant, such as acting as an RICS assessor for the Assessment of Professional Competence, or receiving payment for speaking at a CPD event.

If a surveyor has been made redundant, the principal redundancy payment, up to a maximum of £30,000 is not normally taxable (as income tax or national insurance). Amounts in respect of notice in lieu, and outstanding leave, are subject to income tax and national insurance. A redundancy payment above £30,000, and any other payments, add to the income for the year, and could mean that the tax payer/surveyor is pushed into the higher rate/40% bracket, even if their regular salary is lower. Surveyors facing redundancy should seek advice from an accountant on the implications of the timing and amount of redundancy payments, and try to negotiate suitably tax efficient arrangements with their employer.

If a surveyor is self-employed, the same rates of income tax apply as with salaried employment, except that they are applied to the profit earned in the accounting period/financial year. Whereas income tax (and national insurance — see p 55) is due for the salaried surveyor at the point income is received, self-employed surveyors and partners typically pay by instalments on account in January and July each year (see p 142).

In the case of a limited company established by the surveyor, income tax is payable on a salary in the same way as described above with salaried employment. Any additional sources of income are accounted for as part of the end of year self-assessment tax returns in the usual way. Surveyors' tax liability could also be influenced by capital gains tax, such as when shares or an investment property are sold (and there could, in fact, be losses which help reduce overall tax liability).

Despite a sole trader's possible cash flow difficulties, self-employment provides a better cash flow than salaried employment from a taxation perspective. This is especially true in the early stages of the business — ie income tax and national insurance is paid later than the point at which income has been received (in contrast with monthly deductions with salaried employment). The ability to earn interest on income received, but not yet due as taxation, creates a further, albeit small, advantage. As indicated in Chapter 10, VAT can be managed to create cash flow benefits.

Interest

Bank and building society interest is usually received net of tax at 20%. For example, £10,000 savings earning £400 during the year, as per the bank statement, sees interest credited at £320, and tax deducted at 20% (ie £80). At the end of the tax year, the bank or building society forward a certificate to the tax payer/surveyor stating the interest earned and tax deducted. The tax payer has no further liability for income tax if total income for the year is below the higher 40% rate band (as above, £35,115 — 2003–04), but any income above £35,115 would be taxed at 40%, with a further 20% being due to the Inland Revenue (total tax of £160 — ie 40% of £400). The point at which the interest is paid determines the tax year in which the surveyor has liability for any income tax (as opposed to the interest being apportioned between years).

The surveyor deals with any additional liability for income tax through filing a self-assessment tax return. The Inland Revenue may not automatically send a form to salaried employees/surveyors whose income is below the higher rate band (which would usually act as a prompt for payment), but whose bank interest takes into the higher rate band. Consequently, there are surveyors in salaried employment not making the necessary returns, but who because of the receipt of interest, do in fact have a further liability for income tax which should be declared.

The taxation of bank/building society interest at the higher rate is one reason why a surveyor in the higher rate tax band could arrange for savings (and/or other investments) to be held in the name of a husband or wife who is a lower rate tax payer. In the case of a non-working husband or wife receiving no other income, use would be made of the £4,615 personal allowance (2003–04), and the 20% initially deducted by the bank would be recoverable (as it would in part if falling in the personal allowance and 10% and 22% bands). Also, investments in joint names generally mean apportioning the income in an equal share, but there are situations where income needs to be treated differently because of an unequal share in an asset. Advantage should also be taken of cash ISAs/TESSAs which are free from income and capital gains tax liability.

Dividends

If dividends have been received from equity/share investments, the company pays a dividend, for example, of £9,000. However, this counts as income of £10,000 gross to the tax payer, with the £1,000 representing the tax payer's 10% income tax liability (£1,000 ÷ £10,000). The £1,000 is known as a "tax credit".

The basic rate tax payer, whose total income for the year does not exceed £35,115 (2003–04), has no further income tax liability (and the 10% is not actually a liability — see calculations, p 58). The higher rate tax payer becomes liable for dividends at 32.5%, and has a further 22.5% liability (and again the 10% is not actually a liability). It is worth noting that as the further 22.5% is based on the gross amount, the actual rate of tax on the dividend element (which represents the profits drawn from the business) is 25% — ie 22.5% of £10,000 (£2,250) is 25% of £9,000.

As with the comments in respect of bank interest, surveyors in salaried employment could be liable for income tax on dividends from shares, having not been prompted to make a tax return. Unlike the tax deducted by the bank/ building society on interest by a low income earner, the 10% tax credit on dividends is not recoverable.

Property income, and rent a room relief

A surveyor with property investments, such as a second home, or small portfolio, has to account for investment income. Allowable expenses are deductible, such as letting and managing agents' fees; legal and accountancy fees; water rates, council tax, service charges (depending on whether the rent is inclusive or exclusive of such items); any rent paid; and an allowance for wear and tear (which is generally 10% of the rent — or alternatively is based on the actual cost of renewal). It is therefore the profit which is subject to tax in the same way as a self-employed surveyor/small business would set costs against income. Whereas a surveyor's own home/principal residence would not be subject to capital gains tax, capital gains with an investment property would be subject to capital gains tax.

If surveyors are considering acquiring investment property, guidance on income tax requirements is available from the Inland Revenue, and the assistance of an accountant is likely to be required. Where surveyors receive income as trading income rather than investment income, different accounting and taxation rules apply. Very broadly the distinction is due to the nature of the activity, frequency of acquisition, trading, etc.

If a surveyor rents out a room in their own home (whether owned freehold, long leasehold or as a tenant), £4,250 (2003–04) is tax free. This is known as rent a room relief. If the income is below £4,250, a tax return does not need to be completed. However if a tax return is made, a declaration needs to be made if income has been received (but not the amount). If the rent is above £4,250, the income is declared on the tax return, with allowable expenses being deducted if appropriate. If income is received by other people for other parts of the home (such as joint owners), the tax free allowance is adjusted (and the Inland Revenue is available for guidance if this is the case). Surveyors renting out a room should also check with mortgage providers and insurers whether there are any implications.

National insurance

National insurance liability differs between self-employed status, and receiving income as a salaried employee. The latter is still relevant to a surveyor/director receiving a salary through a limited company, or paying salaries to staff under any form of business.

Salaried employees

Salaried staff have to pay national insurance in addition to income tax, and employers additionally have to pay national insurance contributions for their members of staff. These are known as "class 1" contributions.

National insurance is calculated on a weekly basis for salaried staff, albeit deducted when the salary is paid, such as on a four weekly or calendar monthly basis.

For the 2003–04 tax year, employees' class 1 contributions are calculated as 11% of the weekly earnings between £89 and £595, plus 1% of weekly earnings above £595. The additional 1% of employees' contributions commenced from 6 April 2003 — a further illustration of how tax increases are made more subtly than raising the basic and higher rates of income tax.

Employers' class 1 contributions are calculated at 12.8% of weekly earnings above £89. Unlike employees' contributions, there is no upper earnings limit. Where an employer provides other relevant benefits for an employee (such as a car or private medical insurance), employers' class 1A contributions are payable at 12.8% of the benefits.

For the 2004–05 tax year, employees' class 1 contributions, are calculated as 11% of the weekly earnings between £91 and £610, and 1% on weekly earnings above £610. Employers' class 1 contributions are calculated at 12.8% of weekly earnings above £91 (with no upper earnings limit).

Employers account for national insurance themselves, or use the services of an accountant. The Inland Revenue produces guidance on the deductions necessary for income tax and national insurance, and the returns that need to be made.

Self-employment

A self-employed surveyor is liable for class 2 national insurance contributions and class 4 national insurance contributions.

Class 2 contributions for 2004–05 are £2.05 per week, payable at £26.65 per quarter, totalling £106.60 per year. Class 2 contributions are payable once the surveyor has notified the Inland Revenue that they have become self-employed, and commence from the date that they become self-employed.

Class 4 contributions are calculated on an annual basis in line with the self-assessment income tax system, and are based on the profits of the surveyor/ business. For the 2003–04 tax year, 8% is paid on profits and gains between £4,615 and £30,940, plus 1% on profits and gains above £30,940. For 2004–05, 8% is paid on profits and gains between £4,745 and £31,720, and 1% of profits and gains above £31,720.

Class 2 and 4 contributions are the same in respect of a partnership, paid by individual partners, with class 4 contributions being based on the partners' individual share of profits.

Comparisons between self-employed and salaried situations

To compare the respective liability between a self-employed surveyor and a salaried surveyor using the above example where £25,000 (£480pw) was paid (and which is equivalent to profits), national insurance would be as follows (working to the 2003–04 rates, and ignoring class 2 contributions for the self-employed surveyor).

		£	
Salaried:	Employee	2,236	11% of £391pw (£480 less £89) × 52 weeks
	Employer	2,602	12.8% of £391pw (£480 less £89) × 52 weeks
		4,838	
Self-employed:		1,630.80	8% of £20,385 (£25,000 less £4,615)

The self-employed surveyor's class 4 national insurance contributions would be exactly £1,630.80. The salaried surveyor's aggregate liability throughout the year could differ slightly, as calculation is on a weekly basis, and the total salary could have been distributed irregularly. The above figures are based on 52 weeks, and are also rounded. Income tax liability would be the same, as calculated on p 52, at £4,249.50.

As a further illustration of the cost of national insurance and the merits of different company structures, etc, the following national insurance calculations are based on a salary/profits of £40,000 (which equates to £769 per week), this time working to the 2004–05 rates.

		£	
Salaried:	Employee	2,968	11% of £519pw (£610 less £91) × 52 weeks
		83	1% of £159pw (£769 less £610) × 52 weeks
	Employer	4,512	12.8% of £678pw (£769 less £91) × 52 weeks
		7,563	

	£	
Self-employed:	2,158	8% of £26,975 (£31,720 less £4,745)
	83	1% of £8,280 (£40,000 less £31,720)
	2,241	

Income tax liability would be the same, as calculated on p 52, at £8,207.60.

The calculations show that if considering employing staff, employers' contributions add considerably to the cost. It also shows that if a surveyor setting up a practice is paid a salary through a limited company structure, employers' contributions similarly add to the cost. Between the salaried and self-employed scenarios above, at £40,000 salary/profits, the total national insurance for salaried/company arrangement is around three times as much as for the self-employed arrangement. The same points are relevant to partnerships. However, the net distinction is not quite as great, as the company sets employers' national insurance against income/tax in the same way as salary/other costs.

Corporation tax

Corporation tax is payable on the profits of a limited company.

The full rate of corporation tax is 30% of profits. This applies to both the 2003/04 and 2004–05 tax years, with tax years running from 1 April to 31 March (not 6 April to 5 April as with income tax).

The small companies rate is 19% of profits, with the small companies limit being £300,000 profit. If profit is between £50,000 and £300,000, corporation tax is simply calculated at 19%.

In 2003–04, profits up to £10,000 were taxed at 0%. Following the 2004 Budget and Finance Act, if profits are below £10,000 and retained in the company, corporation tax is still 0%, but if dividends are distributed, corporation tax is payable at 19% on the dividend.

For profits between £10,000 and £50,000, a sliding scale applies, taking account of the 0% corporation tax liability. Corporation tax for £40,000 profit would be £7,125 — calculated by £40,000 × 19% = £7,600, less marginal relief of £475 (£50,000 − £40,000 = £10,000 × 19/400). This is 17.8% (£7,125 ÷ £40,000). For £30,000 profit, corporation tax would be £4,750 — calculated by £30,000 × 19% = £5,700, less marginal relief of £950 (£50,000 − £30,000 = £20,000 × 19/400). This is 15.8%. The 19/400 (19 ÷ 400) is simply a formula used. If surveyors form a limited company, accountants will undertake all the calculations.

The figures assume that profits are not distributed. If profits are distributed, the calculations change, and generally become more complicated. Dividends are, of course, paid from the amount available after corporation tax has been deducted from profits, but dividends could still be paid above this amount, such as because profits have been retained in previous years.

If all of the net of tax profit is distributed from a company with profits of £10,000 to £50,000, corporation tax is a straight 19%. If the profits of a surveying practice were at this level, most of the post-tax profit is likely to be distributed, and it is only where significant profits up to £10,000 are retained that the rate of corporation tax reduces — although the liability remains when dividends are eventually taken. The actual distribution of dividends does not affect corporation tax liability for reasons other than the distribution rule.

As mentioned on p 54, when dividends are distributed, the basic rate tax payer has no further liability, and the higher rate tax payer has a further liability of 22.5%. However, as also indicated, these rates are not based on the dividend payment, and if, for example, £75,000 dividend payment was received from post-tax profits, the tax payer's income tax liability would actually be based on £83,333 (£75,000 dividend, £8,333 tax credit, £83,333 gross).

Examples of profitability and tax efficiency

To illustrate how options in respect of business status influence profitability, examples are provided below of a surveyor:

(a) working as a sole trader and paying income tax and national insurance on profits through self-employment
(b) operating as a limited company and withdrawing dividends
(c) operating as a limited company and drawing a salary.

The calculations are based on profit of £100,000, and work to the 2004–05 tax rates. It is assumed that tax payer has no other income.

Sole trader/self-employment

Income tax: £32,207, calculated as £4,745 at 0%, £2,020 at 10% (£202), £29,380 at 22% (£6,463), £63,855 at 40% (£25,542).
NI (class 4): £2,840, calculated as 8% of £26,975 (£31,720 — £4,745) = £2,158 plus 1% of £68,280 (£100,000 — £31,720) = £682. *NI (class 2)*: approx £106.
Total tax: £35,153, equating to 35% of £100,000. Marginal rate (ie on further income) is 41%.

Company — dividends

Corporation tax: £19,000 (at 19%), leaving £81,000 post-tax profit (and dividend), £9,000 tax credit and £90,000 gross.
Income tax: £12,117, based on £90,000, calculated as £36,145 at 0%, £53,855 at 22.5%.
Total tax: £31,117, equating to 31% of £100,000. Marginal rate is 39%.

Company — salary

Breakdown: Salary £89,191, employers' NI £10,809 (total £100,000).
Income tax: £27,883 calculated similarly to self-employed above.
Employees' NI: £3,541, calculated similarly to self-employed above, but at 11% plus 1%.
Employers' NI: £10,809, calculated as 12.8% of £84,446 (£89,191 – £4,745).
Total tax: £42,233, equating to 42% of £100,000. Marginal rate is 48%.

Self-employment and limited company status can also be compared on the basis of profit of £35,000.

Self-employed

Income tax £6,413, NI (class 4) £2,190, NI (class 2) £106. Total £8,709, equating to 25% of £35,000.

Company — dividends

Corporation tax is £6,650, at 19% of £35,000 (assuming profits are distributed and there is no further income tax liability on dividends).

As an example of tax planning, a salary of £4,745 would reduce profits of £35,000 to £30,255 and corporation tax to £5,748 while incurring no income tax or NI, and equating to 16% of £35,000. A salary could similarly be paid in the "Company — dividends" example on p 58, and reduce tax by approximately £2,000.

Timing of dividends and tax liability

One benefit of limited company status is that the payments of dividends, and therefore income tax liability in relation to the higher rate income bands, is controllable in terms of timing and the tax year in which income falls. If a particularly high level of profits are likely to be earned over a few years, but not on a longer term basis, the limited company/dividend structure helps shelter profits from the 40% income tax plus 1% national insurance level, and enable dividends to be paid out more progressively — and in line with keeping income below the 40% higher rate band. Likewise, as indicated in Chapter 7, stages of growth cause variation in profitability. However, as the above example illustrates, a limited company and dividend arrangement is not especially advantageous in the case of relatively high profitability where all the profits (after tax) are distributed as dividends to an individual in the higher rate tax bracket.

Retaining profits for expansion

With limited companies, profits are also more efficiently retained in the business — ie as they are subject to 19% corporation tax, rather than a sole trader suffering 40% income tax and 1% national insurance, total 41%, once into the higher rate band (and a lower rate tax payer is still paying 22% income tax plus 8% national insurance, total 30%). For some types of business, such tax savings are effectively a source of funding, and minimises the need for loans, thus reducing finance costs. For a business with cash holdings, the retained profit earns interest (although business accounts do not tend to provide a rate of interest quite as high as domestic accounts — but at least tax on income is initially at the 19% corporation tax rate instead of up to 40% income tax for the higher rate tax payer). In contrast, a self-employed surveyor working to the self-assessment arrangements (and likewise for partners), is taxed on the year's profits in line with the 0%, 10%, 22% and 40% rates. The tax payer does, of course, need to be able to afford to leave money in the company to secure tax efficiencies. Potential CGT liability also needs to be considered, bearing in mind that retained profits/cash balances add to the value of the company if ever sold.

Husband and wife, and other arrangements

Where a husband or wife works alongside the surveyor, tax efficiencies may be achievable by paying dividends to both husband and wife, benefiting from one or both parties being in the lower tax band (and not suffering the additional 22.5% on dividends in the higher tax band). This should be favourable to a husband or wife being paid a salary — by a self-employed surveyor as well as a limited company (as basic rate income tax is due at 22%, and both employers' and employees' national insurance is payable). Salaries and employers' national insurance are however costs set against profit, which therefore reduces tax liability. It is important to take an accountant's advice on any such husband and wife arrangements, and not fall foul of Inland Revenue rules. There could be other people able to work for the business and receive dividends, although again, accountant's advice should be taken. It is important that any arrangements with family and others are genuine, rather than simply devices to save tax. The Inland Revenue keenly scrutinises such situations.

Combining arrangements

Within the options outlined above, surveyors could operate a limited company as well as receiving income separately as a sole trader through self-assessment. Flexibility to pay salary through a limited company (instead of only dividends) was mentioned above, and this could apply to both husband and wife. The above illustrations assumed that there was one principal. Calculations could also be undertaken where there are partners/other shareholders. As well as a number of surveyors constituting a partnership, a husband and wife business could be a partnership. Notwithstanding the number of possible variations, it is important to avoid tax affairs becoming over-complicated. The tax rules also change.

Investment income

Investment income would also be relevant. If savings accounts and shares were in the name of the surveyor's husband or wife who would not reach the higher rate band, liability at the higher rate could be avoided. (Joint savings accounts require interest to be shared equally for income tax purposes, and the completion of self-assessment returns.)

As an example of the finer opportunities deriving from advanced thought and planning, if the surveyor had building society savings, and, say, £5,000 interest was likely to be received on the building society's usual annual date of 1 June, this would fall in the 2004–05 tax year (as the tax liability is in the tax year when income is received). For the surveyor's business which is expanding into the higher rate bracket for 2004–05, the tax liability on the interest could be 40%, equating to £2,000. The interest would ideally be charged to the 2003–04 year at only 20%, as the total income remains under the 40% bracket. This could be achieved by closing the account in the 2003/04 year, as interest would be paid on closure rather than at June in the 2004–05 tax year. Bank and building society accounts providing monthly interest would also have avoided interest accumulating, and being subject to 40% tax because of its June payment in the 2004–05 year.

Statutory compliance, and costs

With limited companies, statutory compliance and company accounts requirements are more extensive and costly compared with self-employment/self-assessment. Significant additional costs are only, however, accountant's fees, and if the surveyor is well organised with financial records and undertakes basic Companies House duties personally, any extra costs should be limited. Accountant's fees are an expense which can be set against income/profit and reduce tax.

Benefits of National Insurance

Although as indicated in the examples below, national insurance contributions (other than class 2) could be avoided through a limited company where dividends are paid, surveyors might pay national insurance in order to gain earnings-related pension benefits, and any other national insurance benefits (albeit limited for most chartered surveyors, eg job seekers allowance). However, a salary of between £87 and £89 per week (2003–04) preserves such benefits, while not being high enough to be liable for income tax and national insurance.

Self-employed or employed

As another illustration of the rules, it is not possible to be classed as "self-employed" instead of "employed" as a convenient way of securing tax efficiencies. A husband or wife could not, for example, be classed as self-employed if simply receiving payment from their partner's business. Instead, there needs to be a bona fide independent, albeit still self-employed, business — ie having a number of clients. If a business pays someone as self-employed instead of as an employee, the business/employer could be liable for the tax and national insurance due.

IR35 rules

Similarly, the Inland Revenue's IR35 rules prevent people using limited company status to reduce their tax liability when they are for all intents and purposes an employee. This is known as being a "personal service company". Again, a genuine business is needed, having a number of clients.

Working with accountants

This section has illustrated the complexities in the tax system, but also shows how particular tax efficiencies could be achieved. It is important for surveyors to work closely with an accountant, and take an overall view on tax issues — including investment income and capital gains tax, as well as the likely profitability and prospects for the business. In practice, business status, and tax planning relates specifically to the surveyor's circumstances.

Chapter 11, Accounting, provides examples of the accounts of a surveyor/small business, together with further issues in respect of accountancy and taxation.

Winning Business

A business which is genuinely conceived from scratch has to work hard to establish itself. In contrast, a business which commences with a client base from a previous employer, or surveyors who retain consultancy work for an employer or other contacts, gain a relatively easy start to their venture. It will nevertheless be looking to grow, and marketing and business development initiatives are still important.

Winning business as a sole trader is also different from the surveyor's usual previous work in winning business for an established practice already enjoying a strong brand and client base.

The early stages of a new business often involve the need to chase instructions, appear eager to work for prospective clients, and deal with rejections. Not all self-employed surveyors feel at ease with such a new role, but just tough it out.

This chapter concentrates on the opportunities available in respect of the marketing of a surveying business which is starting up. As with much of the business literature and theories on marketing and business development, this can appear one-dimensional. In practice, the principal of a new surveying business has to integrate marketing opportunities with all the key drivers in the business — and embrace a theme of "total business management". Chapter 1, when assessing viability, considered issues such as market analysis, potential clients, how business could be won, the need for a mix of clients, the scope to roll out services, cash flow issues and the defensiveness of certain fee lines. As good as marketing personnel are at helping fire the imagination of their clients/colleagues, and producing brochures, they do not always understand the dynamics of a business, nor the industry issues it faces. This means that the surveyor has the primary role in driving the direction and profitability of the business through effective marketing and business development initiatives. It is still important though to draw on marketing advisers, albeit in a secondary, rather than lead, capacity.

As mentioned in the preface, while *Starting and Developing a Surveying Business* aims to provide practical guidance, it is unable to venture into potentially contentious areas such as the effect of age and gender on business dealings, the deployment of particularly aggressive business development techniques, and opportunities which push, if not break, ethical boundaries. In practice, surveyors find additional opportunities, and possibly also constraints, to those detailed below.

Investment needs — marketing

Although expenditure requires careful control, especially in the early stages of the business, self-employed surveyors/small businesses need to be realistic about the level of investment needed for marketing/business development. There is a tendency for sole traders to be reluctant to spend money, particularly when the return from doing so is not directly identifiable. In contrast, there are also sole traders, whose vision, belief, and confidence in direction and development of the business is such that they freely incur expenditure on marketing and business development — including lavish launches and other forms of client hospitality. The approach, for example, of three older surveyors establishing a partnership and looking to pick up institutional clients in the city of London, does of course vary considerably from the approach of a young surveyor setting out as a sole trader in their home town.

Company name and other issues

A suitable company name is needed, although the name of the surveyor could be adopted, with the addition of "Chartered Surveyor/s", "property consultant/s" or "commercial property consultancy". RICS rules regarding names are set out on p 112.

Business cards and a letterhead are required, and need to convey the right image. If a home telephone is used, it does not have to be a dedicated business line, but it needs a business-based message on the answerphone, with the family primed to respond appropriately if picking up calls. If the business name is different from the surveyor's name, anybody trying to contact the business via Directory Enquiries is unable to do so unless the company name is listed.

Mobile phone contact is expected by clients, and a suitable e-mail address is required. This needs to convey the impression of a serious business, as opposed to the amateur impression created by free of charge hotmail and freeserve type addresses.

Website presence is increasingly important, although not necessarily essential, depending on the nature of the business and its clients. A company brochure is helpful, and more detailed summaries of services can be prepared separately, or by way of insert sheets. Template report formats enable business proposals to be sent out to clients, tailored to each client/proposal as appropriate. Surveyors also need a CV style summary of their achievements, track record, etc, which is available to clients.

Unique selling points

Consideration needs to be given to the strengths of the business in terms of key points which should attract prospective clients. Examples include the background and profile of the surveyor/partners, the range of services available and the specialist nature of the services.

The right style and location of offices presents a good impression to clients, and a photograph of premises of particular character could be incorporated into a company brochure and/or website. Where a surveyor works from home, such a profile is not achievable, but promotional advantage could still be gained by selling the personal service which is now available from an experienced surveyor, who is also in a position to charge competitive fees owing to the low overheads enjoyed by the business.

Making use of contacts and other outlets

A surveyor establishing a business in a particular line of work is likely to have a range of contacts to whom details are sent of the new business, and the services available. Depending on how keenly the surveyor wishes to win business, arrangements could be offered to contacts or other businesses in respect of fees for the introduction of business and the general referring of clients. Family, friends could also help generate leads, and perhaps old colleagues from university. Mutually beneficial arrangements could be established with accountants and solicitors.

Opportunities to notify surveyors of the new business are available in *Estates Gazette*'s "Property Life — People" section and similar outlets. Contacts, other surveying firms and local businesses, could be sent a letter. Advertising space could announce the new business in a property journal or local newspaper. In a small town, the local paper may even be prepared to write a feature on the new business and the background of the surveyor. An entry could be made in *Estates Gazette Directory* (the monthly supplement to the *Estates Gazette* journal which lists surveying practices against geographical location, and also specialist areas of practice).

Targeting clients and client groups

The approach to marketing does, of course, depend on the nature of the services provided. As indicated in Chapter 1, some sole traders offer a range of general practice services, whereas others provide relatively specialist services. A general practice business usually concentrates on a local market, whereas a specialist surveyor typically seeks a wider presence.

Chapter 1 explained the opportunities for surveyors to differentiate their services from competitors, and also that a "jack of all trades" surveyor providing too many services risks clients considering that there is a lack of special expertise in the area for which instructions are being considered.

Surveyors should determine who their prospective clients might be (as best as this is possible, even if only broad client groups). Research undertaken when establishing the nature of the business should have developed knowledge of the markets in which services are to be marketed, and any typical target clients. Direct approaches could be made to target clients. Libraries, websites, and directories, help provide details of target groups/clients.

In terms of general marketing, surveyors should be aware of competitors' marketing strategies. As the business establishes itself, it is helpful to refer to other clients of particular stature (ie a client list), and possibly provide case examples of success achieved on behalf of clients.

Examples of the many other ways that surveyors find to develop particular fee lines include monitoring proposals for new roads in order to offer services in respect of compulsory purchase compensation, and monitoring planning applications with a view to providing development/construction services.

Site boards

To let and for sale boards help advertise the surveyor's business, even to the extent that a low value transaction is taken on as a loss leader where a building occupies a prominent location. A small amount of agency work therefore helps secure a profile in a local market, and non-agency instructions to be won — some of which would not be picked up without the presence of boards (and the client's advanced awareness of the brand).

New small practices sometimes approach local businesses to see if there is vacant accommodation which they could try to let, and this raises awareness of the presence of the practice in the market. Surveyors looking to pick up new clients for agency instructions are typically prepared to undertake an initial free of charge marketing report as part of a business pitch.

Boards on run down buildings, and boards which have decayed after being erected for some time (and the property not let or sold), however create a poor image. Fly boarding is where boards are just erected on anyone's property, such as the grounds of blocks of flats, as nobody knows which individual property the board relates to. A "sold" sign is soon added in order to show the prowess of the agent, and if in the meantime enquiries are made, the agent advises that the property is sold and that details are not, unfortunately, available. Chartered surveyors should not engage in this practice.

If involved in agency work, the company's details and logo appearing within adverts for properties to let/for sale in newspapers, property journals, etc, act as an additional marketing tool. Clients are not, however, impressed by surveyors/ firms who market themselves too prominently at the client's expense, nor when the need to advertise in a particular outlet, or repeat advertisements, is questionable.

Exploiting market developments

Developments in practice, such as market changes and new legislation, create opportunities for business leads. In the larger practices, numerous glossy brochures are arranged by marketing personnel, but opportunities to develop instructions in relation to market developments or new legislation, for example, are sometimes not picked up because marketing personnel have insufficient property knowledge, and surveyors are not alert to, or focussed on, such opportunities. In contrast, the surveyor in a small business is better positioned to exploit such market developments.

Aggression in marketing

Compared with other areas of business, the market for surveying services tends to operate relatively placidly, and in a good spirited way. Surveyors from competing firms often act against each other, contact each other for comparable evidence, attend functions and other networking opportunities, and generally need to get on well together.

A surveyor establishing a new business might wish to deploy aggressive, but legitimate, initiatives in winning business, and quickly develop a client base while the level of business earned by existing firms in the market remains relatively static. In many smaller businesses, because personnel are busy, little time is devoted to marketing and business development (and dynamic business growth). Further expansion is not, however, an option which the principals of all small surveying businesses wish to pursue, with the possible changes to existing working methods, other difficulties and added risks (see Chapter 7, Business Growth). In contrast, a self-employed surveyor/new business usually spends a large part of the early stages concentrating on marketing and business development. Even when instructions are won, and the business is off to a good start, an eye needs to continuously be maintained on the potential to expand, and how this is achievable.

When prospective clients have established arrangements in place with competitors, new suppliers/surveyors may rarely be invited to set out their services on offer. Such clients could still be appropriately approached by the newly established surveying practice — perhaps by suitable correspondence and a brochure setting out the services, competitive fees and how they would be pleased to hear from the client in the event that there was anything they might possibly be able to help with (or in the event that services were in due course tendered). Peppering potential clients with telephone calls and other forms of "hard sell" is not appropriate with surveying services, and as indicated above, regard also needs to be given to the benefits of generally getting on with competitors — and not therefore pushing ethical boundaries too far in winning business.

Business development opportunities extend to securing a position on a panel of property advisers used by a certain client, which although not guaranteeing instructions, generally reduces the non-chargeable time incurred in pitching for instructions. Panel appointments are more likely once the practice is fully established, and is able to demonstrate a track record. Clients operating panel systems typically have formal selection criteria and pre-conditions (including a minimum level of PI cover).

Generally raising profile

As well as targeting particular clients, it is beneficial to generally raise the profile of the business. This is particularly important if a range of services is being provided.

Short articles could be written for a local newspaper or trade journal. This is an example of how business development initiatives do not instantly lead to new business, but are highly beneficial for the longer term. There is a tendency for surveyors to expect an advertisement, or article, to lead to phone calls and instructions, and if such leads are not generated, they cease such initiatives. Others devote a suitable marketing budget to maintaining the profile of the practice in the market, and to particular clients, with the knowledge and experience that instructions develop, however haphazardly.

Much advertising in the consumer market is about brand awareness. An television advert for a particular bank, supermarket, washing powder, drink or chocolate bar does not inspire an immediate purchase, but when such goods and services are required, the consumer is more likely to select the previously promoted product than if brand awareness has not been established. Critically, promotions lead to consumers switching brands permanently once having experienced a new supplier's product. Therefore, when clients are seeking surveying services, the surveyors/practices more likely to be approached could be those pro-actively maintaining brand awareness among their potential client base. As well as advertising, the various ways of keeping a brand in existing and potential new clients' minds include calendars, diaries, paperweights, note pads, pens, and many more. It is important, that items are of good quality, and do not infer a low cost operation that lacks quality of service.

The appropriateness of drinks evenings and other hospitality arranged for clients and prospective new clients, depends on the nature of the client business and the preferences of its representatives. Public sector clients, for example, cannot enjoy significant hospitality, whereas in the private sector, advantages are available if looking after certain clients well — indeed, some clients have high expectations that this will be the case. Over-eagerness regarding gifts and hospitality makes other clients uncomfortable, and they prefer advisers to simply get on with the instructions provided.

Conventional advertisements for the business and its range of services could appear in the local press, property press, trade journals, etc. Advertising could draw on the differentiation of services, or niche areas of expertise. Target audiences and their view of an advertisement always need to be judged, and a range of advertisements in terms of design and text could be used. Entries can be made in *Yellow Pages* and other outlets.

Involvement in RICS events, delivering seminars, attendance at events, local sponsorships, involvement with the local Chamber of Commerce all add to profile, as well as generating specific leads. Judgment does, however, have to be made as to the non-chargeable time incurred, its benefits, and how this impacts on profitability. A newsletter could be prepared for clients, and made available more widely if appropriate.

Business presentations

As part of business pitches, particularly for larger clients, surveyors are required to make effective business presentations, both in writing and verbally. For smaller clients, pitches are likely to be conducted through more informal meetings. Even then, the surveyor needs to be well-prepared in terms of the issues to be covered and the questions likely to be asked.

It is important to articulately convey the quality of the service on offer, and demonstrate the track record, and property knowledge, available. Surveyors sometimes have a relatively small amount of time in the presence of prospective clients, and to make a good impact. Good interpersonal skills, and a suitably friendly and professional demeanour are key qualities.

Thought needs to be given to finer factors such as how surveyors are able to best endear themselves to particular types of clients. Some characters alienate themselves from certain groups through being too powerfully dressed and overly well-spoken, but to a client group sharing the same characteristics come across favourably. As mentioned in Chapter 1, in the same way that there is as tendency for employers/managers to appoint people in their own image, clients sometimes do likewise. Business is frequently won because a client sees something in common with, or just likes an adviser. An assessment needs to be made of what drives a client's decision-making, such as cost, quality of service, and enjoyable close working relationships. The sole trader is ideally able to show a special ability to meet such requirements. Trust and loyalty are key, and it naturally takes time for a new business to get established.

Fees

As well as researching fee levels and the bases charged by competitors, and what represents the market norm, fees bases need to be attractive to clients. A fixed fee gives certainty to a client, but if the extent of the work is hard to judge, pricing could be too high and therefore uncompetitive — or the surveyor underestimates the time involved, and the instruction is not profitable. Time based fees may therefore be preferable, and if there is a danger that this appears too indeterminate, and therefore unacceptable to clients, a maximum fee could be agreed. A fixed fee, less any savings, is an alternative, which although still indeterminate, may appeal to the client, and also give the surveyor flexibility. Different bases could be appropriate for different clients. Standard fees such as 1%

of sale price or 10% of a first year's rent on letting may be quoted by other agents, but a new surveying business could pitch at 0.8% or 8% respectively, and sell their expertise in securing deals to a shorter timescale as justification for the fee level.

When fees are incentivised, such as a percentage of the increase in rent achieved at rent review on behalf of a landlord, or reduced from the rent notice figure on behalf of the tenant, care has to be taken that the client does not eventually consider that remuneration was excessive for the task involved, and that the fee basis was misleading (ie because a rent notice stated a figure considerably in excess of market rent).

The fee basis needs to be drafted accurately in the terms of engagement, such as in the case of lettings where a rent-free period is granted, and the 10% fee basis on the first year's rent would apply to the rent following the rent-free period (ie the headline rent which would be higher than the day-one/market rent).

In looking to secure a good start to the new venture, the surveyor is usually content to initially reduce fees in order to win business, and generally develop relationships. It is difficult to raise fees without having explained the approach to the client from the outset, and it is wise to explain that any concessionary fee basis is part of client development initiatives, with an indication of standard fees also being provided. If surveyors wish, they could offer a cut-price or even nil cost initial instruction, expressing how keen they are to show the client the services they are able to provide (while still stating fee rates for the standard range of services available).

One major benefit in establishing a strong profile and track record, is that higher fees are typically be commanded. The surveyor is aiming to win business on the back of both price and profile — ie through price competitiveness/lower fees with some instructions/clients, and greater profile/higher fees with other instructions/clients.

Where clients comprise members the public spending their own, as opposed to their company's money, such as in the case of residential sales and lettings, there is a greater tendency to "chase cheap". To win business, there is a greater need for service standards to be at an acceptable minimum in order that costs are sufficiently low to ensure price competitiveness and profitability.

Clients' appointment decisions

As in the case of established surveying practices seeking new instructions, there are no clear cut ways of winning business. This reflects many factors, including the fickle way that decisions to appoint consultants are sometimes made. Even for a surveying business enjoying a strong established brand and track record in their field of practice, the representatives of prospective clients who are charged with finding consultants could be unaware of such factors, and possibly never have heard of the practice. Decisions could rest on small price differentials, especially if a judgment as to respective levels of service quality cannot be gauged, or if there is a general presumption adopted in respect of the cheapest option. Friends and associates receive favourable treatment, as do firms undertaking good quality work previously.

As an example of the factors inherent in the appointments of clients by their consultants, in the early 1990s the Midlands based surveyor mentioned in Chapter 1, together with the head of a corporate property team, sought a consultant for an approx 50 acre development site, then worth around £75,000 in advisory, planning, etc, work and £50,000 in agency fees. Three of the larger practices were approached. One declined to return a phone call (or was not sufficiently well organised regarding message taking), and meetings took place with the other two. One quickly deployed a team of personnel able to demonstrate a track record, and show how elements of their team, and the firm's expertise, could be drawn on. The pitch and profile was far superior to the other, albeit still capable competitor, and fee levels were similar. The best firm did not however win. Several months previously, their performance in a smaller instruction (a landlord and tenant matter) worth approximately £2,000 fees had disappointed the head of property. On another matter, a graduate had attended a meeting in the absence of the qualified case surveyor — giving the impression of secondary service standards, when the meeting could have been rearranged in the event of unavailability.

Maintaining service standards

The above example shows how service standards, at all levels, are paramount when running surveying businesses. Business should not be lost because of poor quality service, delays in progressing work and poor availability when the client tries to get in touch. Clients of the surveying and other professions sometimes comment on the disproportionate level energy their advisers put into the winning of instructions, compared to the pace and quality of the day to day casework they undertake.

Systems should be in place to contact clients after an instruction has been completed, and establish whether good service was considered to have been provided. It is then established whether clients are satisfied with the services available — and therefore the scope for future instructions. Existing clients are a good source of further business, but care has to be taken not to be too pushy, as this is potentially annoying, and therefore detrimental to the business.

It is important to never take any instructions for granted, and for clients to always see the importance that their adviser places on maintaining high standards of service. Businesses which try to expand too quickly tend to have a greater chance of taking their eye off such service standards, and going backwards instead of forwards. "Complacency can be a killer" and "Don't kill the golden goose". However, sometimes it is preferable to relinquish clients; indeed shaking off poor clients becomes an art, as relationships still need to be preserved for the wider benefit of the business.

If a secretary is employed and/or other staff have contact with clients, it is necessary to observe their performance, and be satisfied that they deploy the same necessary levels professionalism and passion for the business, as they do in their contact with the principals of the business. In property businesses, principals are easily unaware of the poor impression created to clients, and how prospective clients can be deterred by a frustrated, impatient and/or cold, unwelcoming tone adopted by secretaries, receptionists and other staff.

Alertness to competitive threats

When running a small business, it is advisable not to introduce clients to other clients, unless there are compelling reasons. Clients do not need to know who other clients are unless conflicts emerge, although a client list is beneficial for business development (and the surveyor may also wish to provide clients with courtesy updates on other work/clients).

As far as possible, it needs to be ensured that other professional advisers, including competing surveying firms, accountants and solicitors, do not have opportunities to pull work away from the surveyor's existing clients. It would be unwise, for example, to suggest that a client attends one of the mid-sized or a large accountancy practice's free of charge seminar on a particular topic. The client is likely to be supplied with food and drink and approached by strategically designated representatives of the practice to make polite conversation which quickly leads to discussions about their business. Comments follows about possible tax advantages and business opportunities that the current accountants have perhaps not highlighted before, implying the need for higher level support because of the growth of the business (or just to take advice in respect of special services). Relationships established between the client and the accountant could lead to introductions to the accountant's surveying contacts.

In line with the above comments about a new business looking to deploy relatively aggressive business development techniques, existing businesses need to be alert to any competitors working on their clients, or generally deploying effective marketing and business development initiatives. In larger markets with many clients and advisers, the impact is relatively negligible, but in smaller and more local markets, there is scope for an existing firm's profitability to be dented. Competitors' pricing of their services should be monitored on an ongoing basis, and not only established as part of the initial viability assessment. As well as ensuring price competitiveness, this enables advantage to be taken of any scope to increase fees and profitability.

As a general business point, it is important not to open the eyes of others to the level of profitability achievable, and therefore minimise the chance of competition from new market entrants. The larger practices, for example, earning high profits through certain more specialist areas of work, avoid public/market profile of such work so as not to alert competitors (noting also that prospective clients are targeted relatively precisely through tailored individual approaches). Similarly, for the individual surveyor highlighting personal financial achievements, some people are impressed, and reassured as to calibre of their adviser, whereas others are jealous and withdraw from the relationship, thus seeing the surveyor lose instructions.

Maintaining marketing and business development initiatives

This chapter is only able to prompt ideas as to how sole traders can market their services, and develop their business in its early days. Each sole trader's initiatives have specific regard to the nature of the business, the services available, typical clients, and market practice. Some surveyors take a particularly tenacious approach compared with competitors who are less focussed on marketing and business growth, whereas other surveyors operate less aggressively and develop the business more slowly.

As business begins to be won, the surveyor's attention, and available time, increasingly turns to case work — but it is important that business development continues to be actively pursued. Aside from the issue of expansion, busy periods of work could soon come to an end, leaving the surveyor with relatively little work. There is also a tendency for sole traders to become content with being busy, and not develop the full potential of the business because of a natural aversion to the tasks involved in winning business. The sole trader should continually be assessing how new, more profitable work, is to be won, and deploy the required marketing and business development strategies.

One benefit of the business having been established for a few years is that the surveyor is confident that more business can be won if necessary. This helps a more relaxed outlook to be taken of the future of the business, thus relieving the common pressures and stresses experienced in the early days.

Business Growth

If a new surveying business makes a successful start, thoughts soon turn to expansion. This chapter examines the issues associated with expansion by a sole trader, and then provides a more general overview of issues involved with business growth. Although guidance on the general running of a surveying business is outside the scope of *Starting and Developing a Surveying Business*, issues associated with business growth and business management still influence the aspirations of the sole trader, and the strategy formulated at the outset.

Personal aspirations

Many surveyors who establish their own business on the basis of a sole trader, and have not taken on any staff, wish to remain as sole traders and never employ others. Part-time administrative support, however, is sometimes sourced, including through a family member, or someone in the area looking for part-time work.

Whether expansion is considered worthwhile depends on a range of factors, including:

* the surveyor's aspirations regarding profitability and wealth
* the surveyor's lifestyle preferences
* the enjoyment of independence with self-employment
* the surveyor's need for a bigger challenge
* whether the surveyor wishes to chase clients for new business
* the extent to which business life is forced to merge with work life
* whether the extra efforts and stress are worth any increased financial returns
* the wish to avoid the sort of staff issues which caused so much frustration when in regular salaried employment
* the economics and viability of taking on other staff.

Expanding areas of practice

Expansion is possible into other areas of practice, without taking on other staff — helping to enhance profitability and/or add more variety to day to day work. The setting up of a business generates interest and momentum, and sufficient enthusiasm to put in long hours. In due course, the same work week in/week out, becomes more mundane, leading to the surveyor undertaking fewer hours, and being more inclined to seek other leisure interests. In looking to develop into other areas of practice, the business benefits periodically from an injection of fresh inspiration.

As mentioned in Chapter 1, a business model pursued by relatively few people, but often highly successfully by those that do, is to begin in business at a lower level than the surveyor's full potential. This gives an edge over competitors, helps win business, and profitability is established with greater certainty and speed. In due course, the surveyor taking this approach often wishes to develop into more challenging and profitable areas. Businesses work well when a manager has personal experience of the day to day running of the business, and can provide accurate instructions to employees, complemented by ongoing strategic control of the business.

It is important to consider profit margins from particular activities, not simply fee levels. Higher quality/higher level work with better fees, could, for example, involve additional costs, a reduced ratio of chargeable time to non-chargeable time, and not be as profitable as more straightforward, incisively managed, work.

Realising growth potential

Businesses sometimes remain too busy for too long on lower level/lower fee work, and fail to review the success of the business achieved to date, and make moves to develop higher value/higher fee earning work. Even within the large property consultancies, surveyors and departments can perceive themselves to be successful because they are busy. However, while one busy department (nationally or in a particular office/location) maintains annual profitability and modest year-on-year/inflation based increases, retains the same total number of staff for three years, and does not actually grow; another doubles its fee income and profitability over two years, and regularly takes on new staff. The difference lies in the extent to which senior surveyors pro-actively review the performance of their business, and its growth potential — rather than simply address the instructions the business is receiving. The very fact that a department is busy confirms its potential, but, the worst examples of busy but static operations see management preside over a day to day environment of haphazard reaction to instructions and an often artificial urgency, which while adding to the daily buzz means that the business, overall, is actually under-achieving.

Individual surveyors and partnerships should pro-actively take time out, say every three months, to think about how the full potential of the business can be optimised (see 'Reviewing the business plan' below). Similarly, individual surveyors do not always seek to accelerate their own personal development because of a pre-occupation with day to day issues, and the consequent lack of energy devoted to how the necessary sources of further success are put in place, including by concerted commitment to lifelong learning/CPD.

Platforms for progress

As a general point, a way for a self-employed surveyor or small practice to expand is to continually seek platforms for progress. This includes identifying activities that lead to other, better and more profitable, instructions, even if the activities are not necessarily the most profitable work that could be taken on at the time. The surveyor also has to be selective about taking on instructions of low value and/or which do not fit with the wider aspirations of the business. The Midlands based surveyor referred to in Chapter 1 reached over £200,000 profit pa in the third full financial/tax year, still working alone from home without staff. Although a rather extreme example, and representing an excessive number of hours being committed per week, it is an illustration of how concerted dedication in the short term could lead to a subsequent winding down to enjoy increased leisure time thereafter, while still working on selective business interests. The surveyor's success reflected a clear strategy for one opportunity to lead to another. In contrast, surveyors/businesses simply content with being busy, do not look to develop further stages of growth, and miss out on the higher financial rewards available. The range of services, location, markets/client base does, however, simply inhibit growth for individual surveyors/small business in certain markets.

Evaluating the potential for expansion

If, for example, a self-employed surveyor/sole trader is earning an annual profit of £40,000 (£55,000 fees, £15,000 costs) through a substantial commitment to the business, and by working from home with low overheads, the viability of expansion is worked through in line with calculations on fee income, costs (fixed and variable), profit and profit margins, chargeable hours, rates per hour, and other fee bases.

To take on a surveyor with several years' post-qualification experience, a combination of factors could account for a surveyor's decision to remain as a sole trader. They include a salary of £25,000 paid to the new surveyor, possible other employee benefits having to be provided, employers' national insurance contributions of approximately £2,500, increased accounting/administration, time taken on management/liaison/instruction (the principal's lost fee earning time) and the need to perhaps take on a proper office at £5,000 or more per year. Motivation from employees is typically less than that of a principal, fewer hours (including chargeable time) are more likely, holidays are granted (which account for approximately 10% of the working year), and there is a greater tendency for days to be lost to sick leave. Staff with reduced experience/expertise command lower fees, and the principal has less personal contact with clients.

There are recruitment costs, risk of losing staff, risk of staff wishing/intending themselves to become self-employed and in due course being competitors, risk of clients being prized away if staff move or set up alone, risk of staff providing poor service, difficulty in removing sub-standard staff, protective/restrictive employment laws, and other additional issues.

There is now a pressure for the business to exceed, say, £90,000 turnover for the principal to retain the £40,000 level of profit, and cover the costs relating to the above list of points (say, £90,000 turnover, additional costs because of the new surveyor £35,000, total costs £50,000 and profit £40,000).

The surveyor is likely to conclude that such expansion is not worthwhile, and it is preferable to enjoy the steady level of profitability and avoid unnecessary pressures. All the factors that make a good business viable for an individual surveyor, and help give an edge on competitors through lower costs, lower fees and personal service are potentially eroded as the surveyor expands the business, thus diminishing the competitive advantage.

If things did not work out, there are residual costs of office accommodation, and possibly redundancy payments — such factors subsequently acting as a drag on profitability, together with any attendant disruption caused by the failed expansion.

Guarding against competition

Sometimes sole traders consider that expansion is feasible if only they had someone of the same expertise, and dedication as themselves. Such characteristics are likely to mean that anybody taken on has the ability to similarly run their own business, and although having no current aspirations, once having seen how the sole trader works, and the level of profitability achievable, soon wishes to take such a step, and therefore becomes a competitor.

The sole trader is nevertheless able to promote the possibility of the member of staff becoming a partner: indeed, this is a key selling point in recruiting staff, and also retaining their commitment for many years. The partner is commonly required to inject capital to reflect the goodwill already generated by the business, and this supports any financial costs incurred with expansion. A new partner may, however, provide great advantage to the business, including the introduction of clients, which means that capital is not necessary.

Making the most of support staff

There are alternatives to taking on another surveyor, such as recruiting a bright youngster who is an effective administrator, and can take on areas of work which do not require surveying expertise, which otherwise have to be done by a sole trader. A secretary could contribute more widely, and similarly undertake basic property work. This frees the surveyor to work an increased proportion of chargeable hours — which should be at considerably higher rates per hour than the costs incurred in sourcing part-time support.

For the surveyor working from home, secretarial/administrative support need not warrant the taking on of premises, and costs are likely to be confined to a small salary, especially as only part-time support is needed. Using the above figures, a £5,000–£7,000 salary/total cost helps the surveyor expand the business, say to £80,000 fees and total £25,000 costs (fixed costs remain the same, but there are some variable/additional costs incurred in line with increased turnover), thus increasing profitability from £40,000 to £55,000.

The taking on of a graduate is another option, which, although less expensive in terms of salary, involves time in instruction and training, and still leaves the sole trader undertaking much of the administration (unless there is part-time secretarial/administrative support). An older, semi-retired, surveyor could prove beneficial to the business, possibly on a part-time basis. Linking up with other small businesses is unlikely to be ideal where they are effectively competitors, but where services are complementary or separate (such as a general practice surveyor involved other areas work, a building surveyor, a planning consultant or an accountant or solicitor), there should be advantages, even if only with administrative and secretarial requirements.

Expanding a partnership

In contrast to the initial example above, there are situations where the successful development of a new business creates compelling reasons to take on other surveyors. A business of three partners in their 50s, as illustrated in Chapter 1, with one shared secretary and an administrative assistant situated in city of London offices, and offering niche investment consultancy/agency services is likely to find this the case. The financial profile of the business could, for example, be total fee income of £600,000, profits of £400,000 and £200,000 fee income per surveyor — which might sound high, but calculates at about £120 per hour for 35 chargeable hours over 47 working weeks (net of holidays). Fee income could in practice be higher, especially if agency transactions and other non-time based work have been successful and raised the overall rate per hour — and also if there is a higher number of hours committed.

Profit margins are 66% (£400,000 ÷ £600,000), and costs include premises having the right profile, equipment/cars, staff, marketing and entertainment/hospitality (but not partners' remuneration). The surveyors are earning £133,000 each. The practice is doing well because of the profile of its partners and the niche consultancy services they offer.

Two other surveyors with five to 10 years' post qualification experience could be taken on, and accommodated in the same office, drawing on the same secretary and administrative assistant. If their extra salary package totalled, say £80,000, and each new surveyor secured £125,000 additional fees each, the additional profit of £170,000 would be distributed to partners at £56,000 each, thus increasing their current £133,000 each to £189,000. However, part of the partners' time/fee earning capacity is now lost to non-chargeable management time. On the other hand, work can be passed to new recruits which helps optimise the partners' use of time.

Stages of growth and costs incurred

The above examples show that some elements of expansion carry additional non-staff costs, whereas other stages of expansion are accommodated within the existing structure of the business, and incur minimal, if any, additional non-staff costs (although an additional chair, desk, and computer may be needed, if not additional office space or secretarial or administrative support).

Further stages of expansion involve more substantial costs, such as a new office, relocation costs, and additional administrative support. However, whereas the necessity for secretarial and administrative support often makes a considerable difference to the profitability of the sole trader, a £15,000–£20,000 package in relation to the above partners' individual fee earning capacity of £133,000 is not particularly significant. At this level, such support is essential, not least because of expectations from clients, the need for phones to always be answered and certain duties quickly performed. In contrast, the sole trader has more scope to run the business in a way that means immediate personal availability is not expected by clients, nor is the opportunity to speak to a secretary (although a message facility is, of course, essential). However, those small traders who present the impression of a larger venture than is really the case, cannot afford to be caught out through being difficult to contact and the obvious absence of other staff.

Other elements of cost become necessary with expansion, that are not incurred in the case of a sole trader. As a simple illustration, it now becomes inappropriate for the principal to run a vacuum cleaner round the office and clean toilets, and secretaries usually consider this outside their duties — so cleaners are required. Newspapers are needed for a larger reception area, and coffee and water machines are acquired. A franking machine replaces the principal's own purchase and licking of stamps. File storage requirements increase, and storage/archives become a property cost. Invoicing and credit control is easily managed by the sole trader, but other staff are needed with a larger venture. Computer backup systems are likely to be required, and systems put in place to guard against negligence claims.

Although, in theory, the principal's time is freed from administrative and other simple tasks, and this allows greater time to devote to fee earning, in practice, the administrative and other tasks may have been undertaken by the principal in time which was really additional to the working week (such as an evening or Saturday morning when fee earning work would not have been undertaken). The simple administrative tasks also facilitate gentle thinking time which is beneficial to the business. The extra costs may therefore be genuinely additional costs, not off-set by increased fee earning time (although dependent on the extent to which administrative and other tasks take time in the working week).

Medium and longer term gains

For some businesses, even though the initial profitability deriving from expansion is marginal, profits come through strongly in two or more years' time. This is similar to the beginning of a business where time is taken to develop a client base, and for instructions to be won — and therefore for profitability and positive cash flow to emerge. Another surveyor taken on by a sole trader is unlikely to have an immediately high workload, but could be very busy in six months' time, and the business is then looking for a third surveyor. Because there are investment costs in a business, whether to start up initially or to expand subsequently (including loss of fee earning time as well as financial costs), profitability might appear relatively low. However, once the business is running at full potential, fees are far higher, costs reduce and profits are greater. Business planning, vision and confidence is essential in order to realise such growth potential.

The larger the business, the lower the profit margins are likely to become, but the higher the level of fee levels that could be commanded (with fees, for example, reflecting the greater expertise and improved profile, better clients and higher value work). As illustrated in Chapter 1, the tighter the margins, the greater sensitivity of profitability to changes in fee income and costs, and the greater the proportion of fixed costs, the even greater the sensitivity of income changes on profitability.

When working on scenarios and figures similar to the above, and considering margins, it is necessary to reflect factors such as the principals'/partners' remuneration being regarded as a profit rather than a salary/cost, thus creating higher margins than would be the case where numerous salaried staff were taken on.

People management

People management is a key part of business acumen, and is necessary in order to get the best out of people, and optimise profitability. It is similar to one football manager's team consistently finishing high in the league with ordinary players, while another manager's team fights relegation despite having considerably better players and greater financial resources (ie over-achievement against under-achievement). The self-employed surveyor without significant managerial responsibility previously, still needs to ensure that the management style acts as a key driver in business success — rather than as a hindrance, as is sometimes the case in established property businesses, large and small.

Motivation

In the same way that the principals need sources of motivation, inspiration, variety and esteem, so do staff. For a surveyor/partnership, people management becomes a particular issue as more staff are taken on, and principals increasingly lose day to day contact with staff. Various business texts and theories are available on people management, but its implementation in practice is remarkably simple in a small business — and often reflects the fact that a principal cares and thinks about such issues.

Setting examples and creating cultures

If managers arrive late and leave early, the imperative of hard work in the eyes of staff risks being lost, and it would be better to suitably engineer early starts, late finishes and other days on site/at home to cement the hardworking ethos in the minds of staff. Any excess remuneration earned by principles is ideally not known to all staff, and particularly expensive cars do not necessarily create the right impression. Similarly, a culture of disciplined cost control is important.

A Christmas party is an effective way to demonstrate the value of staff to the firm, including funding the event (and selecting a Friday to ensure a good and memorable time is had without causing problems the next day). In contrast, some firms require staff to arrange and fund events themselves (sometimes in the evening rather than work time), seek a £5 contribution from staff or ask if they wish to have the money instead.

Staff social activities help team building, but it may be preferable for some staff not to become too close and talkative with one another. Wild nights out could carry wider adverse consequences, as well as a hectic social scene being a distraction through excessive office chatter instead of work activity. Principals need to avoid compromising themselves through excessive drinking and any other indiscretions, and also ensure that adverse reputations are not created (and which damage respect from colleagues).

Principals could ensure that they are aware of staff birthdays, lead the signing of a card, and also buy a gift. Flexible hours are attractive to staff, although the office is ideally covered by telephone answering from early morning to early evening. Answerphones just before 9am, after 5pm and at lunchtime create a poor impression to clients and also lose prospective clients. Dress-down Fridays are popular for some staff, but to others weaken a work-hard/play-hard ethos, and are seen as an irrelevant gimmick.

Salaries and staff contentment

Financial rewards are also important for staff, although money is not everything, and people need to feel challenged, inspired, recognised, and rewarded. Although people do not necessarily work for the level of money itself, it is an intrinsic measure of value and achievement.

It is important to gauge the contentment of staff, and consider whether they are adequately valued and rewarded through their level of financial remuneration. Staff becoming disgruntled and feeling undervalued, do not always say anything to managers — rather instead find employment elsewhere, with a manager's late attempts to increase remuneration being to no avail once the member of staff is settled on their change in circumstances. Some staff though are simply greedy — always conscious of the earnings of others, and pointing to advertisements of jobs which pay more. Staff are also refused pay increases only to leave and replacements require the same or higher pay (in addition to recruitment costs, lost fee earning time, and disruption to the business and the client's interests). It is sometimes, however, preferable to allow certain staff to drift away. Even for the occasional good member of staff who departs, many others remain on only modest salaries, and thus have a beneficial impact on the profitability of the business. Reasons for remaining often include the enjoyment of the working environment, including encouragement and reward from managers, and the combined working and social relationships established with colleagues. Again, it is managerial talent which is key in achieving this.

Good and bad managers

Good managers tend to inspire loyalty, whereas bad managers are more likely to be reluctantly tolerated by their employees until an alternative option arises for them. Good managers give constructive criticism, whereas bad managers criticise habitually Encouragement, acknowledgement of achievement, recognition of contribution, and teambuilding all raise the esteem, commitment, productivity, and profitability of staff — as well as gaining respect, which in turn, enables the occasional necessary harsh few words to be met positively by staff.

There needs to openness regarding mistakes, and principals made aware of complaints, possible adverse publicity, potential PI claims, and also sometimes apparently innocuous client queries regarding fees/invoicing. Principals are also likely to wish to lead new business pitches, and personally be involved in quoting terms of business. As with many of the points made within this chapter, as the business develops, more delegation is needed, and the running of the business increases in complexity. Other managers similarly need motivating. Bonus systems may need to be in place, and certain costs of the business apportioned appropriately between surveyors/departments.

Despite the theories on good and bad management techniques, sometimes an unpopular and belligerent, yet very commercially driven, manager delivers a firm greater profitability than the more likeable manager with better people and managerial skills. This is, in fact, one reason why apparently poor managers sometimes continue being promoted within property consultancies.

It is also interesting to see how surveyors and also other professionals/staff deal with the expansion of a business. Some adopt a "captain of industry" aura, and a more macho approach, revelling in the chance to demonstrate ruthlessness — reflecting their vague impressions as to the characteristic hard-nosed commercial culture of big business. In contrast, successful and wealthy business people appear surprisingly relaxed in their day to day work, belying the actual tenacity they apply to the further development of their interests.

Building teams

Many aspects of good management are not particularly challenging to identify, but just need to be afforded due attention. Good management practice needs to permeate throughout the business, and managers within the business who are unaware of their adverse traits have to be suitably orientated towards the required managerial style. Good managers are also able to assemble the right team of people, having complementary skills and the ability to work as a team (noting again, the example of a football team at the start of this chapter, where the right blend of players, together with team spirit, often results in achievement beyond expectations).

Cost control — and total business management

Cost control becomes an increasingly important issue in a growing business, particularly as people other than the principals have scope to incur costs, and principals are less able to monitor expenditure personally.

The principal's detailed knowledge of the business helps determine the costs which can reasonably be cut, and those which are key contributors to profitability and should remain. A danger of cost cutting is that it takes away elements which serve as key lines of investment in client development and fee income. Cost cutting then serves to contract the business, notwithstanding the apparently beneficial short term cost reductions and commensurate increases in immediate end-year profitability.

As a general business point, when things are measurable, such as the amount, in pounds, of costs cut, there is a greater tendency for costs to be cut — but when things are not measurable (such as the actual returns from expenditure/ investment) the investment is not as likely to be made. Knowledge of the business and its markets, together with an element of vision, is key. A particular danger to a surveying business, as with other businesses, is an accountant seeking to cut costs without understanding the business, or adequate liaison with people who do — making a name for themselves through short term cost reductions and profit increases, but causing medium to long term damage by contracting the base of the business.

Similarly, business development and marketing personnel are able identify elements of a property business which are less profitable than others (and seemingly could be discontinued or receive less resources), but do not always understand, for example, how certain apparently lower profit work leads the profile of the business, and is essential in securing higher profit work. Also, the services offered by a surveying business often reflect the need for some defence against market downturn, and the business does not forever pile in resources to the most immediately profitable line of business. Total business management means that principals understand all elements of the business, and involve themselves in areas which influence the growth, direction and profitability of the business. Despite the above comments, a business could have incrementally added to costs over the years, which even though right to incur at the time, now have scope to be cut or eliminated.

While improvements should always be sought within businesses, another general example of business issues is the tendency for new personnel to sometimes wish to change things to make a more noticeable personal impact, and a point of reference for future measures of success. Principals need to keep a close eye on the attitude and approach of new recruits, and ensure that such changes are beneficial for the overall business.

Short term focus

One element of business, ranging from the running of a small surveying practice to institutional investment trends, is the extent to which short termism predominates. On the one hand, the presence (and even artificial engineering by management) of short term pressures and targets helps create adrenalin and momentum within the business and for its individual personnel which should add to productivity and profitability. On the other hand, it means that relatively little focus is given to the more strategic aspects which help the business grow to be bigger and better over time.

The measure of success for a business and the financial rewards for its individuals is often skewed to immediate and year end profitability. It is likely to be beneficial for the business to engender a dual ethos of high week by week/ month by month performance, and longer term vision, growth aspirations and planning. The time, energy and cost devoted to strategic analysis and longer term business planning is actually a minimal element in overall activity, but in being afforded due attention and disclipline, serves the business well in terms of year on year growth.

The various small ways to bring about the right culture include the incorporation of individuals' efforts regarding business development, business growth, etc, into performance appraisal and remuneration. Some people comfortably work for, say, 10 hours a day on things that need doing, but are unable to work seven hours a day on things that need doing, plus three on things that do not — illustrating the need for management to force attention on matters of longer term importance. If, however, managers' rewards are also short-term in focus, the problems persist.

Another benefit of the longer term vision and a growth culture is the scope for staff to visualise the greater financial rewards and managerial positions that are likely to derive in due course, compared with alternative positions of employment elsewhere. This aids retention, and facilitates greater contentment with current salary levels. As an example of finer issues, some surveyors in larger practices operating as partnerships are termed "partners" despite not having a stake in the business as equity partner. The award of partner status raises the new surveyor's esteem, and commitment to the business (and also facilitates higher charge out rates to clients).

Notwithstanding the above comments, a small business sometimes expands very quickly, and at a desired pace, on the back of the work it wins in the market, clients introducing other clients, clients providing more work to the business and modest marketing efforts. In this case, concerted business development efforts are perhaps unnecessary. Attention still, however, needs to be given to the direction of the business, whether the right type of work is being won and whether there are new areas to exploit.

Reviewing the business plan

As mentioned in Chapter 6, the business plan should be reviewed regularly, and supplemental business plans prepared. In the same way that the initial nature and direction of the business was determined in detail, similar processes are necessary periodically. This includes the revised formulation of strategy.

As mentioned in Chapter 2, principals' thinking time is a key driver in developing a business (as it is in surveyors' personal development). Similarly, the time taken out to analyse, re-direct and invest helps reap disproportionately greater rewards than if looking only to deal with day to day case work (ie looking to calmly secure tomorrow's pounds rather than frantically chasing today's pennies). Also, an undue pre-occupation with day to day issues risks a lack of alertness to other/external issues. This is why someone who is detached from a business, and is looking in, including in the wider market/industry context, easily sees things that go unnoticed by the firm itself.

At three-monthly reviews, principals, partners and members of any team/ department could each be asked to identify five areas of opportunity to improve the business. More specifically, each could be asked for a range of points, such as:

(a)   three key areas of growth into new markets
(b)   who are the best and most profitable clients and what is the scope to win more business from them
(c)   three key areas of cost saving.

As well as senior managers, younger surveyors and administrative staff/secretaries are an important part of such a process, and are able to provide an objective view, as well as contributing on detailed areas of the business in which senior managers are unlikely to have day to day knowledge. This is also a useful way for senior managers to show that the business is a team effort, and that everyone's input is valued.

Senior managers have the opportunity to outline the plans for the business, and how individual team members are able to best contribute. In larger teams, the exercise is an opportunity to find out what everyone actually does, enabling the business to present more of a united front to its market and its clients — and which sometimes leads to more business being won (even if accidentally rather than any concerted attempt at cross-selling).

It is important that the principals seek appropriate advice, and do not assume that all expertise rests with them. Business growth could be a new concept, notwithstanding the established nature of the business. In Chapter 1 it was commented that, "The instincts of highly successful business people are hard to explain, and difficult for others to understand. They just 'seem to know'." Even when some individuals are provided with advice from highly successful business people/consultants, for example, it is still ignored, especially where involving elements of vision, understanding, belief.

In the case of family businesses, there are particular issues to consider, such as succession planning, the roles and responsibilities of individuals, and dealing with family members who lack the necessarily expertise or commitment.

Keeping close to clients, competitors, and staff

As always, principals need to retain their focus, being alert to any threats to the business, and be pro-active regarding new lines of opportunity. Major companies still take their eye off the ball, and become remote from customers, and suffer as a consequence (Marks & Spencer, for example). It is therefore important to remain close to customers/clients as well as to staff.

An eye also needs to be maintained on competitors. There could be a burst of business development activity, and a new business/sole trader may be looking to aggressively win business in a local market. The business itself, in theory, should always be looking to develop itself with reasonable aggression, but in practice, its principals' preferences, or a sensible period of more progressive and cautious growth, means that there is a relatively relaxed commercial attitude.

Work undertaken by staff needs to be monitored to see whether anyone has too much freedom to cultivate certain clients, perhaps with a view to setting up their own practice, joining a competitor or even moving to work for a client. Staff contracts need to be suitably protective.

Planning for stages of growth

It is shown from the above that different stages of growth give rise to particular issues. The many advantages and disadvantages of growth need to be evaluated. Businesses that seek to grow too quickly risk running into difficulties, although if well-managed, rapid growth is an exhilarating and profitable experience for a business and particularly its principals. Surveys have, however, shown that rapid growth often creates discomfort for staff and increases illness and absence — owing to constant change, high workload, new staff not being taken on quickly enough and perceived greater pressures — rather than fear of job security. With rapid growth, attention is easily diverted from all the usual facets of good people management.

The business needs to remain alert to the various economies and diseconomies of scale associated with growth. Although, as mentioned above, profit margins fall as a business moves from its early roots (such as an individual sole trader or small partnership) to employing salaried staff, as long as overall profitability is increasing sufficiently, the principals reap greater personal rewards. The business could provide an increasing range of services to the market, and it is important to identify those which are susceptible to market downturn (such as agency and development/planning work) and those which are relatively protected (or "defensive") from downturn, such as estate/property management. Such issues are similar to those outlined in chapter 1 when assessing the viability of setting up the business, and how a suitable range of services might be determined. Many of the principles which applied when establishing the business initially continue to be relevant, albeit on a larger scale. The business will have new key drivers which can be identified — really the principal factors upon which success turns.

Legal and accountancy issues become more complex as a business becomes larger, and it is important to ensure that appointed advisers have the necessary expertise. Advisers who sufficed in the early stages of the business may now be inadequate (and expanding businesses outgrow their advisers). They could still however be retained for some elements, such as a local accountant for payroll and VAT, and a larger practice for corporate affairs, year end accounts and tax affairs.

Human resources and other issues

The expansion of the business increases the range of human resources issues requiring attention. Recruitment processes need implementing, and/or formalising, and employment policies establishing. This includes holiday entitlements, and policies on absence and sick pay. The ability for people (particularly malingerers) to still be paid for being on sick leave and not working is one of the aspects that often rankles with owners of businesses — as does other areas of employment law, especially as self-employment brings no such benefits. Maternity and paternity issues, rights to flexible hours, discrimination and other laws are all factors that the business needs to be alert to, and it has to be accepted that the most commercially favourable decisions sometimes carry other consequences. Certain employment law and health and safety requirements kick in once a certain number of people are employed, and it needs to be ensured that staff/managers, as well as principals, observe the relevant regulations. Pension arrangements may have to be put in place.

Poorly performing staff might have to be removed, and advanced planning is necessary to guard against a range of difficulties. Systems need to be in place to determine the activity and profitability of staff, such as time sheets and case lists. Performance appraisal systems are helpful. Exit interviews as well as regular liaison with staff helps the business improve. If good staff are to be secured, a small business needs to ensure that it is attractive to staff. Office location, working environment, local amenities and accessibility are important — and general practice surveyors should be well-placed to consider such factors in line with property overheads and lease issues.

Outsourcing options could be appropriate for certain services, and this includes staff through agencies as an alternative to part-time and full-time arrangements administered through PAYE. Temporary/agency staff can be called in at short notice if secretaries are ill or on leave, and using a range of suppliers, where feasible, helps ensure availability, and also prices to be regularly compared. Contingency planning and risk assessment is necessary. The need for businesses to evaluate worst case scenario cannot be overemphasised.

Information on employee rights and other employment issues is available under "Employing people" on the Business Link website, *www.businesslink.gov.uk*.

Future issues

If the business develops, further issues become relevant, such as corporate governance, succession planning, the acquisition or merger of other businesses, and disposing of the business or parts of the business.

The goodwill generated, and the capital value of the business able to be realised on disposal, is a potential benefit of being self-employed — and the success of the business should not be judged in relation to annual income and profits. Some businesses are not, however, easily tradable, such as a sole trader with a small number of key clients providing instructions because of the track record, profile and expertise of the surveyor. Others are more saleable, such as a profitable small practice having a number of clients, sold to a larger property consultancy for whom the sole trader/partners continues to work on a consultancy basis (full-time, part-time, permanently or temporarily).

If acquiring a business, the owners are likely to have dressed it for its day, and ensured that profit profiles and other factors help achieve the best price — similar to investment profiles being worked up and even manipulated in order to secure the highest price on a property sale.

Sources of success

For some sole traders, success comes remarkably easily and quickly, and the business provides good financial rewards, together with an attractive professional and personal lifestyle. For others, it takes time to develop the business, and it is several years before it feels established.

The blend of factors that influence the level of success include:

- entrepreneurial acumen
- an eye for gaps in the market
- creative and analytical thinking
- strategic understanding
- vision
- focus
- the planning of strategies
- financial awareness
- hard work
- commitment
- determination
- thinking time
- thinking beyond today's pennies in order to secure tomorrow's pounds
- being prepared to invest
- technical skills
- interpersonal skills
- being on top of market changes
- managerial qualities
- people management skills
- identifying drivers with the business
- knowing what motivates individuals
- being sceptical and streetwise
- attention to detail

- reviewing strategies
- generally being able to instinctively embrace a multitude of elements in the day to day running and future planning for the business.

As has been illustrated, in terms of the economics of the business, it is not necessarily the pursuit of the highest fee levels per instruction, but the relationship between factors such as profile, fee levels, cost base, profit margin, and profit.

Case Illustrations and Other Issues

This chapter illustrates how a surveying business can be formulated, reflecting issues covered in previous chapters. Although new surveying businesses take various forms, the illustration relates to a surveyor, Derek Chapman, who wished to begin trading from offices, and was keen to expand the business.

Derek, at 43, had previous experience spanning a corporate in-house property team as part of his graduate training plus six years' post-qualification employment, and 12 years' experience in a mid-sized private practice.

Experience and services

Experience included a range of general practice activities which were thought conducive to setting up alone within the central area of a city. He hoped to find a market between the large practices involved in higher level work, and the small practices winning instructions from mainly smaller local businesses and private individuals. The services to be listed by the new business were lettings, sales, acquisitions, rent review, lease renewal, and rating. Derek wished to focus on these areas in order to define the business for marketing benefits, and to avoid the burden of administering client accounts in the early stages of the business. Property management services, including rent collection, and service charge management, were to be established in due course if the business did well.

Family and financial issues

Derek's most recent salary was £40,000 with a £6,400 bonus. Internal prospects for promotion were limited, and the scope to move to another practice did not appeal. Financial circumstances included a house valued at £250,000 to include £70,000 equity, a wife in regular employment in the medical sector at £24,000 per year, and savings of £36,000 (including £15,000 in accumulated TESSA/ISA accounts). A small portfolio of shares totalled £8,000. A pension plan was in place with both previous employers, although the first was limited. The package from the previous employer also included a car, and private healthcare. Outgoings were approximately £12,000 mortgage repayments per year (£1,000 per month), and household, utilities, entertainment and holiday expenditure of approximately £15,000 per year (£1,250 per month). This included supporting two daughters, aged 17 and 18, but this was soon to increase with university fees and related support (approximately £8,000 per daughter per year, net of the income from part-time jobs).

Initial issues, and start-up costs

It was possible to deal with RICS requirements, premises and stationery, prior to leaving the current firm, but only a small amount of work from a client investment company had been established, amounting to approximately half-day per week (although this represented a helpful start). The choice of office premises, just outside the city centre, reflected the importance of securing the right profile for the business, and having regard to expansion plans. Whereas some expenditure would not have been necessary if Derek's plans were to remain purely self-employed, it was considered preferable to incur the costs associated with a larger venture from the outset. This made expansion more easily achievable, and was considered most cost-effective overall (as opposed to relocation), even though it required greater expense/cash flow initially.

The premises comprised a relatively small ground floor shop/office, comprising 25 m²/275 sq ft, but which benefited from large signage and a display area for any properties being marketed. Further office space was taken on the first floor, comprising 42 m²/450 sq ft, with exclusive access/security being achieved via stairs at the rear of the shop/office. Five staff could be accommodated, and property overheads comprised £11,500 rent, £3,000 rates, £2,300 service charge (as the property was multi-let), £500 insurances — total £17,300. Three months' deposit at £2,875 and a quarter's rent in advance of £2,875 totalled an immediate payment of £5,750, which in addition to other amounts paid in advance, together with legal costs, meant that approximately £7,500 was required to begin with. The landlord was not willing to grant a rent-free period, but this enabled Derek to negotiate a break clause after one year of a five-year lease (exercised at three months' notice) to include a penalty of three months' rent in order to minimise costs in the event that the business failed to establish itself. The property was in good condition, but the retail/shop area needed fitting out to meet requirements in respect of brand and profile, including antique/character desk/chair/furniture, high quality carpets and an expensive sofa for the small reception/client area. Only a small amount of expenditure was needed to improve the office to be used by Derek for client meetings, but furniture, pictures and plans, were required to create the right image. The above office costs totalled £8,200. A laptop, office computer, plus mobile phone, printer and other equipment, were purchased for £4,100.

A full time secretary with previous experience in the property sector was recruited, and was able to provide administrative and marketing support and carry out some basic property work. Salary and employers' costs totalled £16,700 per year, equating to £5,000 for the first quarter after allowing for advertising/recruitment costs. As Derek had handed back a company car, £12,000 was incurred on a new car which created the right impression to clients (successful, not lavish). An initial marketing budget was set at £3,500 which included an evening drinks and buffet launch, adverts in the Thursday (property) section of the city-wide daily paper, a basic website and brochures. Stationery, professional indemnity insurance, accountants and other costs came to £2,200.

The above set-up costs totalled £42,500 (including those for the first three months — secretary/recruitment, and rent in advance). Running costs were estimated at £800 per month, comprising mainly property and secretarial costs, but also telephone, and other utility bills. As further costs would relate to client work once undertaken, running costs were estimated at approximately £1,000 per month for the first six months. Costs over six months were therefore in the region of £50,000 — although in terms of cash flow, this represented the worst case scenario of no income being earned during the early months. Cash flow is also, of course, influenced by the repayment of the loan, with shorter term loans involving higher monthly repayments.

Financial planning

In taking a cautious approach, the requirement for initial capital was set at £50,000. An examination of Derek's financial circumstances highlighted that Derek's wife's monthly net salary would not cover household expenditure, even if they moderated their lifestyle and expenditure. However, if they used their savings (other than TESSAs/ISAs) household expenditure could be covered for six months without the need for the business to produce income (assuming a £50,000 loan was secured for the business). It was not appropriate to make withdrawals from TESSA/ISA savings accounts (which provide interest on a tax-free basis) owing to the tax-free income being received, and penalties being imposed for closure.

£70,000 equity was available in the home, providing scope for domestic re-mortgaging, or a security for a commercial loan. As interest rates were historically low, there was a risk of higher mortgage repayments being necessary in the future, and a fixed rate of interest over a three-year period was arranged in order to provide certainty (albeit slightly increasing repayments initially). Savings and investments also provided some fall back security, although the income produced was relatively small (and also the majority was received on an annual basis).

Derek remained with the same mortgage provider/bank, as business finance and bank accounts were to be taken from the same bank. Discussions were held with other banks, so as to ensure that competitive terms were achieved. Derek was aware of the benefits of developing a good relationship with a bank, especially in view of the expansion plans for the business, and a likely later need for further finance.

£50,000 business finance was therefore needed, with a neat division being made between domestic arrangements which could cover household needs, and business finance which would support the business. Derek preferred to operate the business in a self-contained manner — ie being prepared to inject capital from existing resources, but not raising finance domestically through re-mortgaging and other credit lines (such car finance, and a number of easy to secure consumer loans). This enabled a fall-back position to be secured in the event of early difficulties — as opposed to having to seek business finance if the business did not get off to a good start (if indeed business finance could be achieved in light of limited early success). It was unlikely that the bank would not require Derek to put some cash into the new business (although the bank may waive this in exceptional circumstances, such as equity in property considerably exceeding the loan/overdraft — say by three times). In order to secure a commercial loan of £50,000 plus overdraft facility of £10,000 (noting that interest repayments would be an additional monthly cost) a business plan was prepared.

The business plan

The business plan outlined Derek's proposals for the business. Set out below is the principal information, with early information in respect of names and other matters being excluded, and editing/amendments being made, including "attached" information rather than appendix references. Information referred to as "attached" is not reproduced here.

Introduction

This business plan sets out the proposals in respect of (name), a commercial property consultancy being established in (location).

The business has ambitious expansion plans, although as demonstrated below, stages of growth are dependent on meeting targets regarding profitability, and the ability for the business to attract the right new personnel.

The business plan is prepared pursuant to the requirement for both start-up and initial working capital. Subsequent plans will be provided in respect of further expansion, and any related financial requirements.

Nature of the business

The services provided by the business will initially be confined to agency (sales, lettings, acquisitions), landlord and tenant (rent reviews and lease renewals) and rating. This will be primarily in respect of retail, office and industrial premises. A glossary of property terms and explanation of property types and sectors is attached, and a "Summary of services" is set out below.

Proprietor, and background

The proprietor, Derek Chapman 46, has 20 years' experience, spanning agency, landlord and tenant, property management, rating and valuation. This includes eight years for (company name), and 12 years for (company name), a mid-sized practice based in (location). The role in private practice included departmental responsibility for business development.

Derek is a Fellow member of the Royal Institution of Chartered Surveyors (benefiting from the designation FRICS), and is also a qualified arbitrator for rent review. He is married with two children, and lives in (location), approximately four miles from the proposed office. Further details are provided in the attached "Summary of personal and financial circumstances".

Other personnel

Only a secretary has been recruited at this stage. Administrative cover is to be provided by a family member who is familiar with all office systems and procedures.

Interest has been expressed by the proprietor's former surveying colleagues in joining the practice, but while this could contribute to the expansion, such possibilities will be reviewed in due course in line with the volume of work being received in the early stages of the business.

Advisers

Lawyers, (name), and accountants, (name), have been appointed. Support is also being provided from (name), who specialise in business and training advice to companies in the property sector. This has proved helpful in formulating the plans for business, and will continue to provide a check on expansion plans. A statement from (name) in respect of the view of the potential for the business is attached.

Status of the business

The business will operate initially on the basis of sole trader, with a view to considering partnership and limited company options in due course.

Premises

Offices are being acquired at (location) which provide a good profile for the business, as well as accommodating expansion plans without the need to relocate. A copy of the agent's letting particulars are attached.

Market research, and market overview

Existing market knowledge, together with further research found the smaller comparable private practices in the area to be busy, with a number of established practices looking to take on graduates. As the work of the larger and mid-sized practices covers a far wider geographical area, their performance is not a direct reflection on potential to develop a new business in (location). It is however known that the larger and mid-sized practices are busy, further confirming the general strength of the market for commercial property surveying.

Over the past five years, it is known that four new small practices have been established within the city. One has an office in the city centre, whereas the three others are situated in the suburbs. In all four practices, the proprietors have remained the only fee earners, and apart from secretarial and administrative personnel, no surveying staff have been taken on.

From discussions with three of the proprietors, the businesses are clearly at a level where expansion would be feasible if not for the preference to remain working to the current format. It is not known how profitable these practices are, but their ability to sustain business and the lack of aspiration to expand, is considered beneficial for (the subject practice). There is considered to be little risk associated with the development of another small new business in the local market.

As reported in (extract — not reproduced here), the outlook for the local economy is good, and this is reflected in high occupational demand in the main property sectors of retail, offices and industrial. There is considered to be minimal risk of market downturn.

In the event of a severe market downturn in the property industry, transactional activity would slow down, and competition between surveyors in respect of fee levels could intensify. This scenario is considered highly unlikely, but if experienced, the range of services provided by the business, and the flexibility to adjust exposure to certain areas of work, will limit any consequences. As operating costs are relatively low, and redundancies could be made if need be, it is highly unlikely that the business would begin to suffer losses. Suitable bonus/salary structures entered into with staff, would also limit the financial impact of reduced income.

Summary of services

The initial concentration of services will help the business develop its expertise in landlord and tenant, and rating, in particular — and therefore, as a small business, be able to demonstrate expertise to the potential client base. These areas of practice are also consistent with the targeted marketing and business development initiatives which are planned (see below). This includes the scope to win instructions by leading with a service, such as landlord and tenant, with a view to offering more services as the practice expands.

Agency instructions will initially be confined to the city centre, while landlord and tenant, and rating, work will be sought within an approximate 30 mile radius, and will include towns of (names). Agency instructions will concentrate on lettings, as clients are more likely to have repeat instructions, and such clients are relatively easily identified within the market. Sales and acquisition services will, in contrast, be promoted more generally, and fewer instructions are anticipated in the early stages of the business.

In the early stages, valuation instructions will be handled on request, but will not be marketed. As part of the planned expansion, valuation services for the purpose of company accounts will be marketed, but it is likely to be some time before positions on lenders' panels are sought, and valuations services for loan security work is provided.

As part of the expansion plans (see later), a full property management service will be developed (including rent collection, service charge management, assignment, sublettings, dilapidations, etc — see glossary). This will enable the business to provide a composite estate management/asset management service to property owners/investors.

Planning and development expertise is not considered essential, but would be developed if the right senior personnel can be recruited. A residential sales or lettings arm is not planned at this stage, although it may be considered as part of future expansion, and will be dependent on the outlook for the residential market, and related agency income, at the time.

Markets for proposed services

The market for the services is competitive, but there are differences between the market of larger clients, and the market of smaller clients. The smaller client market is particularly price sensitive, and smaller surveying practices tend to be appointed. It is however considered that sufficient instructions will be won owing to the profile and expertise of the proprietor, Derek Chapman. Also, although fee levels appear modest, the relatively greater expertise that Derek will have over other surveyors as part of negotiations for typically lower value cases initially, will significantly condense the time spent. As part of the expansion plans, such work would be taken on by a recently qualified or graduate surveyor.

It is the ability to win business from larger clients in respect of landlord and tenant and rating work, that provides an edge for the business, and a key contributor to profitability over the first year or more. The business is being positioned to compete on both price and service against the larger competitors, while at the same time carrying a superior profile in relation to smaller competitors, who in any event tend not to pitch, or be invited to pitch, for such work.

Instructions have been received already from (name), a regional firm of property investors. This will amount to approximately a half-day per week on landlord and tenant matters on an ongoing basis, although, fortunately, as advanced instructions have been given, work can begin, and amount to two days per week in the first two or three months if need be.

Fees

The attached schedule provides approximate fee bases for agency, landlord and tenant and rating work among larger competitors, mid-sized competitors and smaller competitors. The proposed fees for (the subject practice) are also shown, and which are drawn on in the assessments of profitability (also attached). This represents a cautious level of fees which are considered conducive to business development, and once the practice develops, should see scope for progressive increase.

Costs

A schedule of start-up costs and running costs over the initial six-month period is attached. A separate schedule summarises fixed costs on an ongoing basis, assuming the business does not expand, and illustrations are provided of the variable costs that will differ between the agency, landlord and tenant and rating work being undertaken.

Start up costs total £42,500, including property and secretarial costs over the first three months. Total costs over the first six months are cautiously estimated at £50,000.

Profitability

Illustrations are attached of the fees and costs associated with

(a) a full working week, based on the number of instructions which could be handled, allowing 70% of time to be devoted to casework and 30% to be incurred on business development activity/non-chargeable time
(b) a full working week where only 10% of time is non-chargeable
(c) a scenario where instructions amount to only 50% chargeable time per week.

This latter scenario shows that the business breaks even at approximately 50% chargeable time per week, although this assumes that 75% of agency instructions result in fees. Scenario (a), which is most representative of trading between six months and one year, produces profitability of £38,000 per year and scenario (b), which is considered to be the long term potential, ignoring expansion, is £54,000 per year.

Cash flow

The proposed services of agency, landlord and tenant and rating are disadvantageous in terms of cash flow. This is because properties take time to let or sell, and because landlord and tenant and rating negotiations take time to conclude. (Further detail is provided in the glossary of property terms).

Although the cash flow is particularly problematic in the early stages of the business, the further services which are intended to be developed will ease this. This will include further acquisition work as part of agency (as fees will generally be received far sooner), and property management (which provides regular fees).

The attached cash flow statements assume that income will not be received for six months, and thereafter will be, on average, six months following initial instruction.

Statements are prepared in relation to the profitability scenarios of (a), (b) and (c) above. This shows that even with option (c), where instructions amount to 50% chargeable time per week, cash flow is just positive beyond six months.

Marketing

The attached summary of marketing initiatives, including the budget of £3,500, is designed to maximise initial awareness of the new company to specific target clients, and the local market at large.

This comprises five lunches, three drinks events, website, brochures, one local press advert, introductory letters to prospective clients, former colleagues, etc, Christmas cards for three months' time, and high quality branded diaries in view of the time of year.

The agency instructions can be secured from the same client base as the landlord and tenant work, which will be mainly on behalf of landlords/investors. The rating instructions will typically be from separate clients, as tenants or owner occupiers. The agency work helps create brand awareness, and also maintains an active presence in the occupational markets which benefit landlord and tenant work in particular.

Expansion plans

It is the intention to expand the business, although this is to be done in a cautious and progressive way.

Expansion will initially comprise the development of the current (location) office, to include one senior surveyor and one graduate. The graduate will relieve the secretary of some administrative duties, and family support will also ensure that apart from surveyors'/fee earners' remuneration, only a small amount of additional costs are likely to be incurred with this stage of expansion. As indicated above, the offices have been selected in order to support such plans. The target time frame for this stage of expansion is 12–18 months.

In the attached illustrations of profitability, (d) shows projections on the basis that all three surveyors are working to optimum capacity, and that profitability is in the region of £110,000 pa.

Expansion into other towns and cities in the region will work to a similar format to the development of the initial office, although each new location will have a small established client base owing to the geographical area covered by the initial office. Indeed, as part of the initial marketing initiatives, the clients approached and general advertising of the business gives regard to potential new locations.

The development of the second office will involve the principal of the practice relinquishing day to day work in the initial office, and this will warrant the employment of a fourth surveyor — which helps further the expansion.

As shown in the attached illustrations, start-up and working capital for the second and subsequent offices will be similar to the initial office, although items of cost, and the timing of revenue, are different.

Regional expansion and the recruitment of other surveyors will widen the range of services provided. Regional profile, for example, will help establish an auctions team at some stage, and one of the surveyors interested in joining the business has expertise in the licensed and leisure sector.

As this business plan and financing requirements relate to the initial financial requirements, further information is not provided on expansion plans. Updates are however to be provided to lenders on a six monthly basis, together with an addendum to the business plan which comments on progress to date, and the potential for expansion. Three monthly management accounts, and supporting information, will show the level of instructions being won, and costs being incurred. Assessment of the value of work in progress (ie work not yet invoiced), together with amounts due to creditors will help indicate profitability. Future business plans will also comment on aspects such as human resources, internal accounting, information technology systems, staff training, marketing/branding, and risk management in more detail, reflecting their increased complexity with a larger business, and as the principal will be further delegating responsibility for management and compliance to others.

Financial requirements

A loan of £50,000 is required, repayable over five years, together with an overdraft facility of £10,000.

The summary of the proprietor's financial position is attached, and shows equity in the family home of £70,000 and savings/investments of £44,000. The home is available as security, but savings/investments are ideally retained to provide a fall-back position. £15,000 can, however, be committed to the business if need be, thus reducing the requirement in respect of loan/overdraft.

In the event of failure of the venture, and a return to salaried employment, a salary is achievable in the region of £40,000, although three to six months should be allowed to secure a suitable position.

As indicated in Chapter 4, an executive summary would also be provided at the start of the business plan. A conclusion can be provided at the end.

Analysis of proposals

Derek's attitude is conducive to business success. There is a defined strategy, and Derek's focus is on both the day to day surveying case work, and the growth of the business. Importantly, support is sourced in respect of administrative and other matters, and Derek will minimise any distraction, as well as lost fee earning time on such matters. It is essential that Derek's ability to be creative is allowed to flourish — and is not lost by preoccupation with issues easily handled by others. In contrast to Derek's attitude and approach, surveyors setting up their own business are sometimes averse to incurring expenditure, and opt to muddle their way through various administrative and other tasks, when instead the focus should be on winning business, and undertaking case work.

As mentioned on p 26, thinking time is important, and when Derek is away from the daily workload, naturally occurring thoughts need to relate to the performance of the business and its plans to develop, and not immediate, but largely irrelevant, frustrations with suppliers, utility companies, etc. Derek also wishes the financial position to be comfortable, and not be distracted by frequent monitoring of cash flow, and having to provide regular updates to concerned lenders. Prudence is however taken in securing a loan plus overdraft, rather than simply a loan. Also, if income comes in sooner than expected, this will be due to the success of the business, and will facilitate expansion sooner than planned, such as taking on another surveyor. As expansion reduces cash flow, the full loan facility may still be needed. If partners were attracted however, they are likely to be required to inject their own equity, thus reducing borrowing needs.

Derek's growth plans are ambitious, yet cautious. Disciplined, progressive, stages of growth reassure lenders, and also any future investors/partners that the business is being run well. Although with a larger requirement for finance, growth through new offices could perhaps take place more rapidly, Derek does not see the need for this, nor wish to create undue pressures and risk. He is also conscious of some the points included in Chapter 7, Business Growth — ie the more the business expands, the more remote he will come from its interface with clients, and the more reliant he will be on staff/surveyors/partners to help develop the business in its desired direction. Also, in developing strong profitability in one location, this will support the financial needs to develop another. Derek is also aware of the need to attract surveying and other staff, perhaps including investors/partners, and the opportunities in a "growth business" provide a good selling point.

Derek also shows the ability to understand the economics of the business in terms of cash flow, and profitability. Cash flow statements are simple for surveyors to prepare, and involve stating income and expenditure at relevant points of time, and having a running balance which represents the need for borrowing (almost like preparing a bank statement in advance, to include a running balance). As illustrated above though, it is important to work through certain scenarios, and assess the impact, for example, of income being received one or two months later than planned. In contrast, costs are relatively unlikely to be incurred sooner than planned, or be higher than planned, unless related to more work/potential income, or a greater speed of growth, in which case they are not problematic. Illustrations of basic profit and loss and balance sheet accounting are shown in Chapter 11.

As the bank wished to see the proprietor, as a new business, put some cash into the business, Derek put in £13,000 and secured a loan of £40,000 (at 3.5% above base) plus overdraft facility of £7,000 (at 4% above base). The home was provided as security.

The next three chapters cover three key areas of compliance: RICS requirements, VAT and company accounting/taxation.

RICS Requirements

This chapter initially summarises the support available from RICS, together with its regulatory functions. It then highlights the key points associated with professional indemnity insurance, clients' accounts (Members' Accounts Rules), the need for a complaints procedure and a range of other professional ethics issues which may emerge. The chapter concludes with an outline of CPD, lifelong learning and training — setting out RICS requirements, but also showing how self-employed surveyors are able to cost-effectively keep up to date with market developments. This chapter was confirmed by RICS as being accurate as at June 2004, and there are, of course, changes to the regulations over time.

Research, and RICS registration

Surveyors who have worked previously in private practice should be aware of most of RICS requirements. However, particularly if they have worked for larger practices, surveyors may not be familiar with the finer detail of all the requirements, such as handling clients' accounts, arranging PI cover and ensuring that terms of engagement are in place for a range of instructions. Time is therefore needed to establish RICS requirements, and implement the necessary systems and procedures.

Surveyors who have worked in an in-house role in the public or corporate sector, and have not dealt with external clients, usually have more extensive research and preparation to undertake. In-house surveyors may, however, have experience in observing certain issues and requirements through their appointment of property advisers.

RICS provides an information pack called *Setting up in practice as a partner, director or sole trader*. This is produced by RICS Professional Regulation and Consumer Protection Department, which is responsible for ensuring that members comply with RICS Rules of Conduct.

Guidance is also available on RICS wider requirements. All such sources of information help ensure that the requirements — and also all opportunities and ideas — are being considered.

When making the transition to self employment it is necessary to contact RICS so that the requisite notifications are made, including the name of the practice and contact details.

RICS role, and Rules of Conduct

Under the RICS Royal Charter, the Institution has a duty to act in the public interest, and to protect the interests of consumers/clients. RICS is not a trade body/trade association which pursues an agenda simply orientated to its members' advantage. The RICS Royal Charter and Bye Laws sets the framework for the RICS Rules of Conduct and the obligations placed on surveyors.

A new and updated set of RICS Rules of Conduct were implemented with effect from 1 January 2003, and a small change took place from 1 January 2004. Guidance to Rules of Conduct is an accompanying publication and there are also individual guidance papers produced from time to time, as well as articles in RICS Business. The two main documents are available from RICS in hard copy or CD. The information is also available via RICS website *www.rics.org/resources/standards*.

Professional indemnity insurance

Where RICS members provide surveying services to the public in the capacity of sole practitioners, partners, directors, members of limited liability partnerships and consultants, professional indemnity insurance needs to be held, and the RICS Professional Indemnity Insurance Rules (PIIR) complied with.

It is unlikely that any surveyor setting up a surveying practice does not require professional indemnity insurance. RICS provides a list of "surveying services", known also as "definitions of specialisations" which covers the activities included. As examples, only the entries beginning with "Commercial" are set out below.

Commercial building surveys — Investigating and assessing the construction and condition of a commercial or industrial property. The extent of the survey is agreed between the client and the surveyor. The report includes reference to visible defects and guidance as appropriate on maintenance and remedial measures.

Commercial property agency — Acting as agents for clients wishing to acquire, sell or let commercial property.

Commercial property dispute resolution — Acting as arbitrator or independent expert to determine a dispute, usually relating to the value of a commercial property, or as an expert witness submitting evidence to such an arbitrator or independent expert.

Commercial property management — The management (and sometimes letting) of commercial property, including collection of rents and the calculation and collection of service charges; the specific duties are agreed with the client.

Commercial property valuation — The assessment of the value on a defined basis and stated assumptions of an interest in commercial property. Valuations are required for many purposes.

Commercial real estate management — Full management of the property portfolio to optimise property values and minimise running costs.

Commercial rent reviews (and lease renewals) — Negotiations on behalf of landlords and tenants to agree new rents payable under lease provisions and other terms of new leases on renewal.

Commercial and industrial international property — Surveying practices specialising in any aspect of commercial and international property outside the UK.

If members are involved in work which might not be regarded as surveying services, RICS can be contacted in order to establish whether this constitutes surveying services, and professional indemnity insurance is needed. A retired surveyor undertaking administrative work, for example, albeit in the property sector, is unlikely to need insurance cover.

Where work is undertaken in a consultancy capacity, this means working as a consultant to a firm of surveyors, as opposed to being a consultant/adviser to the usual form of client. Here, as an alternative to the consultant surveyor taking out their own PI cover, they can be named on the firm's insurance policy. (Technically, the phrase is "PII cover", but surveyors usually refer to PI cover.)

RICS members also require PI cover in the event that they are "held out" as having one of the above positions — such as someone who is not technically a partner appearing on letterheads alongside other partners in a way that suggested they are actually a partner.

Level of cover

The level of cover needed depends on the income earned by the surveyor/ practice. If the gross income in the preceding year was less than £100,000, a minimum £250,000 cover is required for each and every claim. For gross income of £100,000–£200,000 in the preceding year, the minimum cover for each claim is £500,000, and if gross income is above £200,000 in the preceding year, the minimum cover for each claim is £1m. The majority of surveyors setting up in practice fall into the first category, although partnerships established by a number of senior surveyors could require a considerably higher level.

It is important for surveyors to establish the precise rules and take advice from an insurance broker and/or RICS. In particular, the cover obtained must be "no less comprehensive than the RICS Professional Indemnity Insurance Policy". This includes covering "every civil liability", which includes breach of contract as well as negligence claims. A "fidelity guarantee clause" has to be included, which means that cover is still provided if losses are due to the fraud of other partners, directors, members (ie of a limited liability partnership), consultants and employees. "The insured" must cover former as well as present partners, and directors, etc.

The policy has to work on an "each and every claim" basis, which means that all claims in a year have to be covered to a minimum of £250,000/£500,000/£1m each (as opposed to "aggregate" cover which covers all claims and acts as a ceiling on total payout within a year).

There is also a maximum uninsured excess permitted by RICS regulations. This depends on the level of insurance obtained: if £500,000 or less, the maximum uninsured excess is 2.5% of the sum insured, or £10,000, whichever is the greater (and if over £500,000, it is 2.5% of the sum insured).

RICS recommends that most members should consider arranging cover of at least £500,000 for each and every claim, and also alerts surveyors to the need to check for any conditions and exclusions which could mean that a policy did not comply with RICS requirements. Again, detailed advice from a broker and/or RICS is available.

As with other forms of insurance, brokers act in an agency capacity and arrange insurance on behalf of the surveyor/business requiring insurance. Surveyors need to ensure that a broker has expertise in arranging insurance for surveyors and understands surveying and RICS requirements, as well as developments in the surveying market and any effect that this has on insurance requirements. If a broker is tied to particular insurers, premiums stand more chance of being uncompetitive, and ideally a broker has access to a range of insurers. Brokers, of course, require a fee — which is either a fee in the same way as surveyors charge clients for their services, or a commission which is included in the premium quoted. It could in fact be a combination of the two. Surveyors could also approach an insurer directly.

RICS operates a listed insurers system, which means that surveyors are required to place insurance with a listed insurer (the current details of which are available from RICS). Surveyors should always be aware of the conditions of any insurance policy, and be sure that they are in compliance. Examples of conditions include the type of work that can be undertaken, and the need to alert insurers in certain circumstances.

Surveyors should obtain a copy of RICS leaflet, *Insurance Claims — Reporting claims and circumstances that might lead to a claim*. This includes guidance on:

- when to notify insurers (including in respect of work which could lead to a claim, and verbal threats to claim)
- how insurers judge such notifications
- how this might affect insurance premiums
- how surveyors avoid compromising their position by saying the wrong things, omitting details from evidence, and being unaware of timescales and other procedural and evidential aspects of law.

Comment is also provided on how larger firms implement systems whereby a designated senior person takes responsibility for PI issues.

Run-off cover

If closing the business or retiring, run-off cover is required. PI cover works on a "claims made" basis, which means that cover is needed for the point at which the claim is made — which could be for a period of up to six years after the surveying services were provided. The insurance policy does not work on the basis that the insurance for a particular year covers claims relating to work undertaken in that year. If a business fails to establish itself, and provides surveying services, but then is considered unviable, six years of insurance cover could add up — although minimised by suitable negotiations with the insurer (and there are also reduced requirements in some cases, reflecting levels of turnover). There may also be legal liability extending beyond six years: six years simply being RICS requirements. Regarding run-off cover, any consultants need to be noted on the policy after ending their period of consultancy for run-off cover to be provided.

If a surveyor/business is unable to obtain cover in the usual way, it may be able to do so from an Assigned Risks Pool facility. This is, however, subject to stringent entry criteria, and there is maximum time of two years. Further details are available from RICS.

This chapter summarises only the basic issues. Full details are contained in RICS *Professional Indemnity Insurance Rules*, available in hard copy from RICS or via the website. For surveyors particularly interested in insurance matters, other guidance booklets are available.

Initial notification to RICS

When initially advising RICS of the details of the business when being set up, confirmation is required that professional indemnity insurance is in place. This has to be provided within 28 days of insurance cover commencing.

A surveyor sometimes wishes to establish their business and notify RICS with a view to then working up business plans, and it may be some time until clients are on board and fee earning work is undertaken. It is not necessary to hold insurance simply because of having set up in business. However, once trading and holding insurance, temporarily not providing surveying services does not escape the liability to hold professional indemnity insurance cover (although in some cases it reduces the requirements, such as if turnover is below a certain level, or particular activities are not being undertaken). RICS also needs to be informed if a surveyor/business ceases to be insured.

Work for friends and family

Surveyors are not able to provide informal advice to friends and family on property matters without having the requisite professional indemnity insurance in place. Whether a fee is received or the advice is provided free of charge is irrelevant. There have been cases where remarks made at social functions, such as those which act as authoritative reassurance as to the wisdom of a property decision/deal, lead to a case of negligence being successfully proved once the property decision/deal proves to have been flawed.

Self-employed surveyors should be insured anyway, and the issue therefore tends to be one for surveyors in regular salaried employment — especially as most family and friends have some form of property interest, and often turn to the known surveyor for guidance. In practice, such requests are easily handled by explaining the lack of expertise in a particular area and/or that it is not possible to give advice, especially as certain, usually extensive, investigations would typically be needed in order for an accurate decision to be made. It is acceptable to advise someone how to source their issue, such as suggesting that they speak to a local firm of chartered surveyors, or that they contact RICS.

Work for clubs and societies, voluntary work and charity work could create a liability if the work is regarded as surveying services. However, if employees obtain from their employer written confirmation that the work falls within (or is deemed to fall within) the scope of their employment, personal liability and an obligation for the employee to arrange insurance should be avoided. Employees could also obtain an undertaking from their club, society, or charity, that action would not be taken against them — the wording of which is available from RICS. Self-employed surveyors/businesses cannot do this with clients.

Low Fee Earners Scheme

For surveyors providing occasional surveying services, and who ought to hold professional indemnity insurance, the RICS Low Fee Earners Scheme may provide a suitable alternative. This would also be applicable to the few surveyors who undertake private work in addition to regular salaried employment. Also, the need to obtain insurance cover may be thought to render any supplementary work unviable owing to the level of insurance premium usually payable, and the scheme may therefore help private activities to become feasible.

The Low Fee Earners Scheme is split into cover which includes surveys and valuation, and cover which excludes surveys and valuation (as such areas tend to be the common areas of negligence claim and liability). Levels of cover are available in bands of annual turnover not exceeding £5,000, £7,500, £10,000 and £20,000 (the maximum turnover to be eligible for the scheme being £20,000). The annual premium is £231, £315, £399 and £720 for the respective bands where survey and valuation work is not undertaken. Where such work is undertaken, the premiums are £294, £399, £504 and £885. The rates are inclusive of 5% Premium Insurance Tax, and slightly higher priced monthly options are available if paying by direct debit. Premiums, as with other expenses of the business, may be set against income/tax.

The policy provides a minimum cover of £250,000 per claim in accordance with RICS minimum insurance requirement. For work excluding survey and valuation, there is an excess of £500 for each claim, and if including survey and valuation, there is similarly an excess of £500 in respect of each claim, but increasing to £2,500 for claims in respect of survey and valuation.

Insurance should be obtainable relatively easily if undertaking a small amount of surveying work alongside salaried employment, or intending to undertake a small amount of work under £20,000 pa as a self-employed surveyor, but there are declarations that have to be made which could hinder insurance being obtained —

such as whether previous claims have been made against the surveyor, and whether another insurer has declined to provide cover. If surveyors are clear of certain declarations having to be made, they include payment with their form, and are covered. As always, false statements and omissions usually invalidate insurance and lead to personal liability — as well as RICS pursuing non-compliance with professional indemnity insurance obligations.

Member Support Service

RICS Member Support Service (MSS) was set up in 2003 following the case of *Merrett* v *Babb*, in which a surveyor was held to be personally liable for negligence as his employer did not have the required professional indemnity insurance. It was, however, an unfortunate situation in that the employer's business had become insolvent and the receiver cancelled the run off-cover. RICS members now, with a few minor exceptions, pay £15 per year to the Member Support Service. This is not insurance, but instead a support fund which helps surveyors in the event that they face claims as employees. This includes access to legal advisers, and help in dealing with insurance providers — although the level of support, including financial assistance, depends on the circumstances involved (such as the conduct of the surveyor in the case in which a claim has arisen and his or her general compliance with RICS Rules of Conduct). A surveyor taking on valuations of specialist properties in which they lack the required experience is perhaps unlikely to receive support in defending a claim.

The service does not provide support to surveyors who should hold professional indemnity insurance in their own right — ie a sole trader, partner or director. Employees also need to be alert to the arrangements that their employers have in place regarding PI cover, and generally ensure that they are not exposed to personal liability/risk.

Clients' accounts

When advising RICS of the details of the business, the surveyor has to confirm whether the business will hold clients' money.

Obvious examples of clients' money include the collection of rents, deposits, service charges, insurance premiums and the holding of clients' money to pay contractors or meet any other expenses/disbursements. Expenses/disbursements met from the company's bank account and later recovered from client accounts is not clients' money, neither are cheques made out directly by a client to supplier, planning authority or arbitrator, and forwarded by the surveyor.

Clients' money needs to be held in a separate account which is clearly identifiable as a client account. A single client account would suffice, but accounts may be opened exclusively for an individual client (known as a discreet client account) if there is a particular need. The money needs to be available on demand to clients, except where alternative instructions are given. The account needs to have the name of the surveyor's firm and the word "client" included in the title. In the case of a discreet client account, the full name of the client is included in the title. Surveyors cannot deduct any fees from clients' money without instruction, nor can payments be made to others (such as expenses, disbursements) without instruction.

Once in every 12 month period, the surveyor needs to send RICS a certificate stating whether or not clients' money was held during the accounting period covered by the certificate. Surveyors have to be aware of certain regulations, even if not holding clients' money — although RICS sends a reminder letter. If money has been held, a report is required from an accountant (suitably qualified as per RICS rules) providing the range of information required by RICS. The accountant is required to undertake test procedures (and this is a cost to the surveyor). RICS can request information from surveyors at any time regarding their accounting of clients' money, and make inspections — including instant spot checks.

Surveyors holding clients' money also have to make a payment to RICS Clients' Money Protection Scheme which ensures that money is available to cover clients' losses in the event of the theft of clients' money by surveyors. This is currently £26 per year.

Surveyors intending to hold clients' money should refer to the detailed guidance available from RICS, as there are many finer points which need to be understood. Details are contained in the Rules of Conduct and Guidance to Rules of Conduct, and other current guidance.

Many surveyors setting up in practice wish to avoid the need to hold clients' accounts and the attendant administrative burden, accounting and cost. Even if handling rents and deposits cheques and also cash could be paid into the client's own account, and direct debits and standing orders entered directly. This could be an account set up by the client for the specific purpose. If the surveyor is not a signatory, it is not possible to withdraw funds — although clients sometimes wish surveyors to have such control. Where a client is able to withdraw money from an account (as opposed to receiving payment/transfer from the surveyor) and the surveyor consequently does not have control of the account, the client needs to be informed that the arrangement is not a client account in accordance with RICS rules, and is not regulated by RICS, nor covered by RICS Clients' Money Protection Scheme.

Complaints procedure

Another requirement when advising RICS that the business is being established is to provide a copy of the business' complaints procedure. A copy of the procedure needs to be available on request to clients and the public. The availability of the complaints procedure should also be stated in terms of engagement entered into with clients.

A model copy is provided by RICS (as below), and is straightforward to adapt in relation to the business being set up.

This note sets out the procedure we will follow in dealing with any client complaint:
We have appointed (*name and contact details*) to deal with complaints. If you have a question or would like to make a complaint, please do not hesitate to contact him/her.

If you have initially made your complaint verbally — whether face-to-face or on the phone — please also make it in writing, addressed to (*name above*).

Once we have received your written complaint, (*name above*) will contact you within seven days. At this stage we will give you our understanding of your case. We will also invite you to make any further comments that you may have in relation to this.

Within twenty-one days of receipt of your written summary, (*name above*) will write to you, to inform you of the outcome of his/her internal investigations into your complaint and to let you know what actions we have taken or will take.

If you are dissatisfied with any aspect of our handling of your complaint and the outcome of our internal investigation, feel free to contact (*name, contact details — eg in the case of a firm, the senior partner, or in the case of a sole principal, another person whether or not locally based, to whom the sole principal is prepared to refer unresolved complaints*), who will personally conduct a separate review of your complaint and contact you within fourteen days to inform you of the conclusion of this review.

If you remain dissatisfied with any aspect of our handling of your complaint and/or separate review, then we can discuss whether we need to go to mediation according to either the Centre for Dispute Resolution (CEDR) or the mediation process run by the Royal Institution of Chartered Surveyors.

If you are still unhappy about the result of any of the above, you can refer your complaint to the "Surveyors Arbitration Scheme" if it falls within the scope of the scheme. This scheme is operated by the Chartered Institute of Arbitrators, Dispute Resolution Services, 12 Bloomsbury Square, London, WC1 2LP from whom you can obtain details.

The example complies with the minimum standards set by RICS and surveyors/ practices may wish to add to the provisions. The time-limits are recommendations to ensure that complaints are dealt with promptly. Mediation is not binding, but in agreeing to go to arbitration, the parties are bound by any decision, subject to a right of appeal. As general practice surveyors will be aware through dealings under the Civil Procedure Rules in respect of dilapidations, lease renewal and other property disputes, the courts encourage settlements out of court, and expect the parties to pursue recognised means of alternative dispute resolution, and not see court as a first option — nor indeed a tactical weapon. The Surveyors Arbitration Scheme is recognised by the courts, and covers complaints, negligence claims and any other claims. Larger and more complex claims may, however, be more appropriately handled by the court.

If a complaint is received, the complainant has to be provided with a copy of the surveyor's/firm's complaints procedure. When the complaint is subsequently made, the stated procedures are then followed.

Sometimes complaints are made directly to RICS about a surveyor/practice, in which case the complainant is asked to fill in RICS standard complaints form. A copy of the completed form is forwarded to the surveyor, provided that the complainant agrees to this. RICS considers the nature of the complaint, and draws the surveyor's attention to the particular Rules of Conduct which have been breached. However, some complaints reflect the complainant's lack of understanding of property matters, and others may be rather petty points of contention.

Depending on the nature of the complaint, surveyors usually inform their solicitor, and perhaps their insurance company as some complaints relate to negligence for which financial redress is being sought.

RICS requires a response from the surveyor, together with supporting information, and this is forwarded to the complainant. If RICS considers there to be no grounds for complaint, both parties are advised of reasons supporting the decision and the file is closed. Where grounds for complaint are established, RICS action could be a warning as to future compliance, or the complaint could be considered by the Professional Conduct Panel — or if particularly serious, a Disciplinary Board. If any surveyors find themselves in this position, the relevant guidance on the processes and sanctions is available from RICS.

RICS itself can deal only with breaches of its rules. Complainants may follow other routes, including redress through the courts. Particularly astute (or mischievous) complainants may adopt a several pronged attack, such as in respect of an agency transaction where the property has been misdescribed and the opportunity misrepresented to a purchaser who finds out the true facts once in the property: civil action under the Misrepresentation Act 1967, criminal action under the Property Misdescriptions Act 1991 (because the complainant has informed Trading Standards) and a complaint to RICS regarding any indiscretions in respect of general conduct.

In 2004, RICS is operating a pilot Surveyor Ombudsman Scheme in Scotland. This involves complaints from clients and the public being handled on an independent basis, with RICS' Independent Appointments Selection Board appointing an ombudsman to resolve a dispute between a surveyor and the consumer. The awareness and use by clients/consumers of such a system reduces the scope for surveyors/practices to deal with a dispute in a way that avoids RICS being aware of the issues. One of the roles of the ombudsman is to report sufficiently serious breaches of the Rules of Conduct to RICS — and which could lead to disciplinary action in the usual way. The scheme could be adopted throughout the UK, and internationally.

The scheme is another example of RICS aiming to be on top of regulatory issues, and ensure consumer protection standards are in place. Such standards also serve as a promotional tool for RICS, surveyors and member firms — creating a perception of integrity and professionalism in the market (although unfortunately often tarnished by reports of activities of estate agents, the majority of whom are not chartered surveyors and regarded by many as not subject to sufficient redress in the event of malpractice).

The commercial reality for surveyors is that it is sometimes preferable to just pay a problem off, especially if the occurrence is relatively rare, and it will not lead to further claims through profiteering. Even with negligence claims and other complaints, although rare, some clients make a speculative attempt to secure settlement, and therefore need handling appropriately.

Breaches of RICS rules

When establishing a business, surveyors are typically focused on winning business and gaining early successes, and consider study of RICS requirements to be a relatively mundane task. This is an example of some of the preparatory work which could be undertaken while in regular salaried employment, thus enabling the business to get off to a focused start.

RICS Professional Regulation and Consumer Protection Department monitors the conduct of firms and individual self-employed chartered surveyors. As an example of the potential severity of non-compliance with RICS Rules of Conduct, *RICS Business* in April 2004 listed a disciplinary board "summary of finding" of "failure to maintain Members' Account Regulations (MAR)", "failure to maintain professional indemnity insurance" and "penalty" of "7 × severe reprimand: 2 × £1,000 fine, 3 × £3,000 fine, 1 × £4,000 fine and 1 × £4,500 fine/undertaking/costs". (Total £19,500).

In addition to fines and costs, if breaches are pursued by RICS, this involves surveyors losing fee earning time, being distracted from the business and receiving adverse publicity through notification in *RICS Business*, on RICS website and also in the local newspaper in the area in which the surveyor/firm operates. A poor impression can be given to clients and business associates, and instructions could be lost. Non-compliance also causes difficulties between partners or directors, and again affects the business considerably (noting that in a small practice, the partners or directors will all be held responsible owing to their obvious direct day to day responsibilities for ensuring compliance).

In the more serious cases, surveyors could be expelled from RICS and no longer be able to practice as a chartered surveyor and use the designations MRICS or FRICS. Disciplinary action could also involve losing a position on RICS panel of arbitrators, independent experts or adjudicators, no longer being an APC assessor or being unable to hold positions in RICS such as through involvement in RICS local branch affairs.

Breaches of the rules do, of course, vary in their severity. For example, on behalf of the Professional Regulation and Consumer Protection Department, RICS Education and Training department works closely with members struggling to meet their CPD/lifelong learning requirements, and disciplinary action is usually a last resort after continued resistance from a member to comply. In contrast, if professional indemnity insurance is not held, or clients' money is not all being held in a client account, action will be taken.

Other issues

This section provides examples of miscellaneous issues relating to self-employment. Many relate also to surveyors working for employers. Some of the issues are specific to RICS, whereas as others are general statutory requirements — many of which RICS still highlight to members through information packs, and articles in RICS Business, etc.

Application of RICS rules

If a surveyor is qualified as a chartered surveyor, and does not wish to actually practice a chartered surveyor, but still undertake surveying services, they do not fall under the RICS rules — and are not obliged to hold professional indemnity insurance, nor comply with any RICS rules. The need to comply with statute is, of course, unchanged.

Anyone is able advertise themselves as a surveyor, and does not need to be a qualified RICS member holding the MRICS or FRICS designations. Other terms could be commercial property surveyor, property consultants, and real estate advisers. Similarly, anyone can advertise themselves as an accountant without being professionally qualified — although to advertise as a solicitor, professional qualification is needed.

Notwithstanding the technical position outlined above, individuals must not hold themselves out to the public/clients as being chartered surveyors if undertaking work in only a surveying/non-RICS capacity. Even if omitting chartered surveyor from business stationary and from the company's name and other details advertised elsewhere, advising clients verbally that the status is, in fact, one of chartered surveyor would still breach RICS rules. Indeed, RICS are likely to scrutinise the affairs of members who are providing surveying services, but working on ways around the rules. The ultimate disciplinary sanction is expulsion from RICS membership.

A risk being taken if practising outside the status of chartered surveyor, is that it is still possible to be sued by clients. If not holding insurance, liability is against the individual's personal assets.

Use of the term chartered surveyor/chartered surveyors

Chartered surveyors will wish to use the designation MRICS or FRICS (or even TechRICS for the technical surveyors). RICS states the following:

> If your organisation practises as surveyors, it has the right to use the designation "chartered surveyors" in conjunction with (but not as part of its name) if the sole principal, or at least half of the partners or directors (in the legal sense), are Fellows and/or professional members, as long as:
>
> - None of the partners or directors has been expelled from RICS in the past, and not been reinstated.
> - An agreement between the partners exists, and has been supplied by RICS, that surveying business carried out by the firm is to be conducted in line with RICS Rules of Conduct; and
> - Where the designation "chartered surveyors" is used in conjunction with a company's trading name, the company's corporate name is included in legible letters on all its stationery, advertisements, signboards and other displays.

RICS adds that:

- When the names of the partners or directors are shown, their RICS designatory letters must be given.
- Organisations qualifying to call themselves "chartered surveyors" may also use in the plural those alternative designations (such as "chartered environmental surveyor"), which the principal or any partner(s) or director(s) has the right to use.

- Do not use additional words after "chartered surveyors" in a way that suggests that other activity undertaken has the chartered status.

RICS also provides the following examples of the acceptable usage (right) and breaches (left):

Joe Bloggs Surveyors Ltd	Joe Bloggs Chartered Surveyors Ltd — is acceptable as a trading name only, not as a corporate name
Joe Bloggs and Associates Chartered Surveyors, Rural Business Consultants	Joe Bloggs and Associates Chartered Surveyors and Rural Business Consultants
Joe Bloggs Limited — Chartered Surveyor	Joe Bloggs Chartered Surveyors Ltd
Joe Bloggs FRICS Chartered Surveyor	Joe Bloggs Ltd FRICS Chartered Surveyor
Joe Bloggs Ltd Chartered Surveyors Estate Agents and Valuers	Joe Bloggs Ltd Chartered Surveyors, Estate Agents and Valuers
Joe Bloggs Ltd Chartered Valuation Surveyors and Estate Agents and Valuers	Joe Bloggs Ltd Chartered Surveyors and Estate Agents
(The only time that firms can link "chartered surveyors" with estate agents under RICS Bye Laws.)	
Joe Bloggs FRICS Chartered Building Surveyor Chartered Facilities Management Surveyor	Joe Bloggs FRICS Chartered Building and Facilities Management Surveyor

The above rules mean that a single surveyor as sole practitioner, with no other surveyors/staff, is able to advertise as chartered surveyors even if this might imply to the market that the business is more substantial than is really the case (chartered surveyors being a sufficiently generic descriptive term — and does not imply that there should be more than one actual surveyor). The designation could be Joe Bloggs Chartered Surveyors (with chartered surveyors, as above, being descriptive rather than being part of the title — and noting that as well as the formats above, the distinction could be conveyed further through the style and size of font or other design).

A partnership providing a range of property consultancy services is not able to designate itself as chartered surveyors if half or more of the partners are not MRICS or FRICS qualified. Commercial property consultants and other terminology could instead be used, and in the case of multi-disciplinary practices that extend services beyond surveying, this more suitably conveys the nature of the business to the market.

Breaches of the above are, of course, dealt with by RICS. RICS cannot unfortunately take action against any non-surveyors/non-members who advertise themselves as chartered surveyors despite not being MRICS or FRICS qualified — or even if they have already been expelled from RICS membership. Instead, RICS works with local Trading Standards offices which bring about the necessary prosecution under the Trade Descriptions Act (possibly drawing also on the Business Names Act).

RICS branding

A surveying business usually draws on branding facilitated by RICS — really as a two-way means of promoting both the surveyor and RICS itself. This includes the use of the RICS logo on letterheads, business cards, compliments slips, brochures, adverts for letting or sale in journals, adverts for the business in journals and directories, websites, site boards, marketing particulars, and outside the business' offices. As well as the usual logo depicting the lion, there is a new strapline which can be used alongside "RICS" of "The mark of property professionalism worldwide".

To see illustrations in practice, and ensure compliance with some of the finer rules on the use of logos, the latest brochure is available from RICS: *Using RICS to promote you and your organisation — changes to the guidelines from 2004* and at *www.rics.org/resources/brand*. This includes a facility to download the logos.

As mentioned in Chapter 5, a sole trader, partnership, limited liability partnership or limited company needs to comply with legal requirements in respect of stationery. RICS also requires a copy of the letterhead when initially informing the Institution that the business is being established.

As an example of both the commercial services operated by RICS and the support to members, RICS Print advises members on printing requirements in respect of letterheads, business cards, and brochures. This includes design templates, and ways in which RICS logos and general branding help enhance the marketability of the firm, as well as complying with RICS Brand guidelines.

Limited companies and Limited Liability Partnerships

Surveyors setting up as a limited company or limited liability partnership, need to include statements in their memorandum of association and articles of association or equivalent constitutional documentation regarding business being conducted in accordance with RICS Bye Laws, and regulations. The wording is available from RICS.

Money laundering

Money laundering is obviously not specific to surveyors running their own businesses, but a sole trader has to take responsibility for the practice's compliance with regulations.

Money laundering involves the conversion of money obtained illegitimately into a form which appears legitimate — and really turning cash assets into paper-based assets which can be sold, with the funds passed more easily through the banking system. Sources include the proceeds of criminal activity, funding of terrorism and tax evasion. Property provides a way of paying in cash to create an apparently legitimate asset.

The rules have recently changed and particularly in view of the fight against terrorism, are likely to change on a relatively frequent basis. RICS produces a guidance note *Protecting against money laundering: A guide for members.* The rules include registering with the Commissioners of Customs and Excise if accepting cash above 15,000 euros (approx. £10,000); checking the identity of clients; record keeping procedures; and informing the authorities regarding cash payments above a certain level and any suspicions of money laundering.

Data Protection

Data protection is another area where the law changes relatively frequently, particularly in view of increased use of internet/e-mail. If personal data is held, the Data Protection Act 1998 provides that details must be registered with the Information Commissioner. Up to date details are available from *www.dataprotection.gov.uk.*

Terms of engagement

For all instructions, surveyors must have terms of engagement in writing, including the fee basis, and how expenses/disbursements are calculated. The drafting of standard letters is another example of the preparatory work undertaken while working in regular salaried employment. Surveyors should be sufficiently familiar with the usual contents. Surveyors who have worked in an in-house capacity should have seen examples from appointed property advisers, and contacts in private practice can hopefully provide samples. If terms of engagement vary during the course of an instruction, this needs to be recorded in writing.

As well as meeting RICS and any legislative requirements, the agreement of instructions in writing minimises the scope for disputes about fees and the services provided, and also support the surveyor's position if having to take legal action for amounts outstanding. Explanations can be included in the terms of engagement as to the detail behind the work to be undertaken, and how fees are calculated. This is, in fact, an opportunity to express to the client the complexity of the work involved, and generally help justify the fees required. When clients do not understand the nature of the work, they sometimes consider that they are being overcharged. Even if clients do not raise the issue, subsequent instructions might not be received, and comments could also be made to other clients. As an example of the need to be accurate, clients should be aware if VAT is payable in addition to the fee. In Chapter 10, it is explained that clients who are not VAT registered, or whose requirements from the surveyor relate to their exempt services/supplies, are not able to recover VAT, and regard it as an additional cost.

Terms can be included which enable the surveyor to retain the interest on any clients' money held, or enable advertising to be off-charged as disbursements at the rate of an individual insertion, even though the surveyor gains a discount for bulk advertising (see p 117). These are included for administrative convenience rather than to profiteer, but it is beneficial to soften such statements, such as by commenting that interest covers bank charges, or additional charges are not made for arranging/designing advertisements. As mentioned above, terms of engagement need to include notification that a copy of the complaints procedure is available (although a separate letter could in fact be provided to this effect). It is possible to work on the basis of terms of engagement initially covering all aspects of work, and instructions thereafter referring back to the initial letter, and containing matters specific to the instruction.

Fees

There are no fee scales advised by RICS, and surveyors are able to charge as much, or as little, as they wish (provided the basis is agreed in writing by the client, and there is no overcharging with time based fees and disbursements).

If surveyors wish, they can undertake an instruction for no, or cut-price, remuneration in order to act as a loss leader for future business. Conversely, they can quote highly in the event that they were keen to be seen willing to work for a client, but did not actually wish to take on the particular instruction on offer.

As good practice, surveyors should quote a fee and stand by the level quoted, and if clients seek to negotiate reductions, this should be resisted. This could be on the lines of the fee being a standard fee, the fee already being reduced from the standard fee, only the usual/reasonable remuneration being sought, and possible reductions being available for a number of instructions taken at once. The general problem with surveyors being prepared to reduce their fees from an initial quote is that the initial quote could be perceived to be profiteering, and that the pricing of professional services should not be associated with bartering. Some clients habitually wish to negotiate, in which case surveyors are aware of this, and consequently tend to initially over-price the fee — thus leaving the client pleased with their efforts, but with no net benefit. Where clients are characteristically poor payers, but are still valued, there is a tendency for higher fees to be quoted (even as a sub-conscious reaction to the frustrations caused by the client regarding cash flow, and payments being chased). Instead of cutting fees at a client's request, the quality of service could be sold to the client, and it is worth checking that the fee quotes have been undertaken on a like for like basis. A different remit would enable a new quote to be made, but there are grey areas between reasonably negotiating fees and fee cutting.

Invoices ideally have sufficient information to reassure the client that the amounts are fair and reasonable (such as through a summary attachment if inappropriate within the invoice description itself).

Surveyors cannot make hidden profits/secret profits — such as charging four individual clients £500 each for a quarter page advertisement in *Estates Gazette*, paying *Estates Gazette* the rate of £1,700 for a full page, and making £300 profit. An exception would be where clients have agreed to this, such as because the terms of engagement state that disbursements in respect of advertising will be charged at the standard rate for the individual insertion and that any discounts due to bulk advertising will not be passed to the client. Other disbursements need to be charged at the amount incurred, and amounts/arrangement charges cannot be added unless agreed. Mileage rates could technically mean that a profit is made (because the rate per mile is above the average cost calculated at the end of the year) but the rate per mile should be stated in the terms of engagement, and is therefore acceptable.

Other RICS issues, and publications

When in regular salaried employment, surveyors would have had access to items such as the *RICS Appraisal and Valuation Standards*, the *RICS Code of Measuring Practice*, the *RICS Manual of Estate Agency Law and Practice* and a range of other guidance notes, publications, and journals.

Surveyors are likely to be aware of the literature they need, but reference can be made to *www.ricsbooks.com* to view the latest information. Estates Gazette Books' website, *www.propertybooks.co.uk*, provides details of other publications which should be helpful. Again, these are further examples of how surveyors prepare for self-employment through preparatory reading while still in regular employment.

To let and for sale signs

If surveyors require to let and for sale signs, these need to comply with the maximum permitted sizes, and the erection of site boards has to comply with the relevant regulations — Town and Country Planning (Control of Advertisements) Regulations 1992.

RICS has a list of recognised site board suppliers that could be contacted. They are aware of the regulations regarding size and sitting of boards, but surveyors will be responsible for ensuring that boards are not, for example, prevented by listed building or conservation area status, or restrictions in a head lease or title.

It should also be ensured that the design of the site board presents the right image for the firm. As outlined in Chapter 6, Winning Business, site boards also help advertise the surveyor's business, even to the extent that a low value transaction might be taken on as a loss leader where a building occupies a prominent location. Boards on run-down buildings, and boards which have decayed after being erected for some time (and the property not let or sold) create a poor image.

Ownership and access to records

For surveyors running their own business, all files and papers generally belong to them, except items which the client has agreed to purchase (such as reports/

valuations, including surveys commissioned from another surveyor/specialist and recovered as disbursements). Copies of leases and other documents also belong to the client.

Parties who legally have access to the business' documents include tenants in limited cases, the obligation of disclosure in civil litigations and rent review arbitration for example, police in connection with investigations including money laundering, Inland Revenue and Customs and Excise.

Clients' confidentiality needs to be protected, and information should not be passed to third parties without the client's permission — unless the third party is legally entitled to the information, as in the above examples. If the police or others are requesting information, legal advice can be taken. Client confidentiality does not extend to money laundering and the authorities must be informed in such cases.

Retained information

In the running of the business, the surveyor should always be conscious of the information that might be helpful in the event of a dispute, however remote the possibility of a dispute appears. This includes being sued for negligence, defence against alleged conduct as part of agency work (such as misdescription or general conduct), a dispute with a client about payment for fees, a difference of opinion with Inland Revenue or Customs and Excise, being able to show that the due procedures were followed when a client was suspected of money laundering — or possibly being able to blame someone else for a problem. It also includes having sufficient information on file to be able to protect clients' interests as part of casework, as again, disputes may arise and extend to legal action, arbitration and other processes where the quality of evidence is likely to have a significant effect on the result. Alertness should also be maintained to information which could prove incriminating, either for the surveyor or the client, and to the extent to which information is committed to computer or paper.

Surveyors using a computer system should ensure that sufficient backup arrangements are in place. Simple arrangements include saving work on floppy disc/CD as well as the hard drive, e-mailing work to another computer/person (such as a family member) and having two or more computers (even a cheap second hand back up). If going away or for other needs, including where work on a laptop could possibly be at risk of theft, surveyors could send an e-mail to themselves with all the information needing to be protected, and then receive it back once logging on-line again. Similarly, two separate e-mail addresses could be held, with one including use as a storage facility. For larger operations, IT support is likely to provide automated backup systems.

As indicated in earlier chapters, businesses need to evaluate worst case scenarios, and where risks are identified, establish how they might be minimised. Worst case scenarios includes loss of the businesses records by fire, and arrangements to separate certain information is ideally in place. Certain files and computerised records are therefore best kept away from the office. It is dangerous to commit particularly sensitive information to computer — another risk relating to theft of the computer.

Destroying information

The space taken by paper-based files increases considerably as the surveyor establishes the business. Roof space or a cellar minimises the problem, but at some point, files still need to be destroyed.

Retention for six years is a generally adopted rule, although there may be no reason to hold certain information for so long as it will not relate to a negligence claim, for example, nor relates to current projects or outstanding fees. Records of income, expenses, etc, need to be retained for six years in accordance with Inland Revenue and Customs and Excise requirements. It is safest to retain information in a case-based yearly sequence, although if there is a particular need to minimise space, accountants' and/or solicitors' advice can be taken. Scanning documents into a computer, microfilming and retention of information on computer hard drive or floppy disc/CD reduces storage requirements, but is time-consuming. Confidential information needs to be shredded or burned.

If surveyors have any particular issues arising in practice, reference can be made to the RICS leaflet, *Whose files are they anyway*. Advice is also available from solicitors and/or accountants.

Alertness to fraudsters

As a general point on the destruction of rubbish, there are fraudsters with systems in place to rifle bins for bank statements and other documents that provide sufficient detail to be able to access the account and obtain funds. Bank statements should be retained anyway, but a letter of innocuous appearance regarding, say, a rise in interest rates would often be binned without much thought, but could contain account details. It is not unknown for business people to seek to obtain the rubbish of another business for whatever motive. E-mail systems are generally unsafe and although problems are unlikely for a self-employed surveyor or small business, people with sufficient expertise may be able to access e-mails and files.

Surveyors running their own business also need to be alert to communications received from fraudsters, and others, whose operations appear to be official, but are scams (some of which are within the law). Examples include letters referring to data protection and other regulations, and how registration is necessary for a fee. Even if the company is providing a genuine service, and registering the surveyor with the appropriate authority, the authority is not the company making the approach, and the charges will be higher than the registration fees payable if going direct to the appropriate authority. Correspondence, pre-recorded telephone messages and faxes all prompt return calls to premium rate phone lines (including high cost overseas tariffs), and care needs to be taken.

Sub-instructing/introducing business

Surveyors are able to introduce business from existing clients and also non-clients to other surveyors (or other professionals) and also take a fee, provided the fee is disclosed to the client/non-client.

Surveyors can also sub-instruct work to other surveyors. There is no obligation to advise the client that the work is being sub-instructed, and the sub-agent is similar to being a consultant/employee. However, unlike the situation where business is introduced, a sub-instruction means that the contract is between the surveyor and the client, and that if anything goes wrong, the surveyor is liable to the client (notwithstanding the liability of the sub-agent/consultant to the surveyor). It would also need to be ensured that the instruction does not breach the professional indemnity insurance requirements and cover, bearing in mind that the need for sub-instruction is often because the work is relatively specialised and outside the usual field of work of the surveyor. Another professional ethics point is that surveyors cannot take on work outside their expertise.

Conflicts of interest

Surveyors need to ensure compliance with conflicts of interest. A conflict of interest is where the surveyor's impartiality is brought into question in view of other relationships/instructions. Examples include not undertaking a rent review for both the landlord and tenant, not undertaking a loan security investment valuation for a client bank when also acting on behalf of the investor acquiring the property, and managing a building in which the surveyor is a tenant (collecting rents, service charges and dealing with disputes).

In some cases conflicts are overcome if clients are satisfied that issues of impartiality do not arise in practice. Where surveyors in salaried employment operated chinese walls and other arrangements, as the sole practitioner, such work is more likely to have to be declined. Conflicts could also arise in connection with the surveyor's interests, such as an interest in purchasing a property, or in relation to group companies (although unlikely to be of relevance to most surveyors setting up in practice).

Where conflicts arise, and it is judged that they can possibly be overcome, the surveyor has to disclose to each client the circumstances, nature and possibility of the conflict, advise each client in writing of the need to take independent advice on the conflict, and inform each client that it is only possible act if both clients confirm in writing that the surveyor can proceed (either on an unconditional basis, or subject to conditions specified by the surveyor and/or requested by the client/clients). In some cases, however, even though the conflict could reasonably be overcome, the surveyor wishes to avoid any question as to loyalty to a valuable retained client when a one-off instruction is available from another party.

RICS Client Referral Service

As part of the initial details provided to RICS, surveyors' areas of practice, and contact details, can be listed as part of RICS Client Referral Service. This may lead to instructions being won when prospective clients enquire with RICS for suitable qualified surveyors who are situated in a particular geographical area.

Managing the business in the surveyor's absence

Surveyors need to consider arrangements in respect of how the business might be managed in the event of long term illness, death, or long holidays. Surveyors are not, however, legally required to nominate someone who will stand in — although under the Financial Services and Markets Act 2000, work counting as an investment business requires a locum.

In the case of a partnership, cover should be relatively straightforward to facilitate within the practice, or through employing temporary staff, but the sole practitioner's business would be particularly exposed without suitable arrangements. A local surveyor or surveyors could be in place to handle the business' affairs. RICS is able to provide more detailed guidance on the aspects needing consideration, including matters to cover in a formal agreement (an example of which is available from RICS), requirements in respect of PI cover, arrangements regarding clients' accounts, access to records, the quality of record keeping, and overseeing the sale of a business.

MRICS or FRICS

Surveyors sometimes consider that their profile is best represented in the market by the designation FRICS, as clients feel reassured by the superior experience and expertise that FRICS is thought to bestow. Details of current requirements are available from RICS.

Updating RICS

In the same way that RICS need to be informed when the business is being established, RICS must also be notified regarding a change in employment circumstances. As well as surveyors running their own business, all surveyors are obliged to keep RICS informed of changes in circumstances.

Conduct

All the usual rules apply to surveyors regarding the way they conduct themselves, including client confidentiality, duty to clients, the need to deal promptly with correspondence. In respect of "Conduct befitting membership of the Institution", the RICS "nine core values" (which are set out to define the professionalism of chartered surveyors and technical surveyors/members) are:

- to act with integrity
- always be honest
- be open and accountable in their dealings
- be accountable for all their actions
- know and act within their limitations
- be objective at all times
- treat others with respect
- set a good example
- have the courage to make a stand.

Also, a surveyor/member (or others on the surveyor's instruction or inducement) should not act in a way which could "compromise or impair the integrity of the member; the reputation of the Institution, the surveying profession or other members; the high standards of professional conduct expected of a member; compliance with any code, standard, or Practice Statement of the Institution or any statute in force at the time; the member's duty to act in the legitimate interest of his client or employer subject to legal or similar constraints; and a person's freedom to instruct a member of his choice". Regarding standards of service, RICS requires that "a member, in the performance of his professional work, the conduct of his practice and the duties of his employment provide the standard of service and competence which the Institution can reasonably expect". As indicated at the start of the chapter, any breaches of such requirements can lead to disciplinary action from RICS, as well as giving rise to financial losses and diminished reputation.

A copy of the rules is available at *www.rics/resources/standards*. Individual areas of practice are, of course, subject to various legislation, and surveyors need to remain up to date on certain aspects. Equal opportunities, and employment legislation also needs to be observed.

CPD and lifelong learning

Lifelong learning is RICS new concept of continuing professional development (CPD).

It means that at all stages of their careers, surveyors evaluate their training needs, and plan activities accordingly. This is in line with personal and business aspirations, and usually includes a combination of technical skills, and business and management skills.

For surveyors developing their own business, research and learning activity is relatively extensive over an initial short period of time. It is therefore worthwhile recording suitable elements of learning when setting up in business — including areas covered in *Starting and Developing a Surveying Business*.

The RICS rules regarding CPD/lifelong learning are set out in the RICS Rules of Conduct. Surveyors need to undertake a minimum of 60 hours of suitable learning activity over any three-year period, with a minimum of 10 hours being undertaken in each year. RICS stresses that it is the quality and relevance of the learning activity which counts, and not only the logging of hours. CPD activity typically comprises courses and seminars, workshop/discussion sessions and private study (incorporating articles, text books, journals, videos and audio tapes).

For surveyors who became professionally qualified after 1 January 2004, CPD has to be recorded via RICS website as part of the RICS CPD on-line recording initiative. This includes the facility to plan learning activities, and RICS hopes that in due course, the web-based system replaces the paper-based means of recording still used by many surveyors.

Self-employed surveyors tend to comfortably exceed the minimum RICS CPD requirements owing to the need to keep abreast of the many developments in law and practice — such as changes to the *Red Book*, new asbestos regulations, the commercial lease code, the impact of Part III of the Disability Discrimination Act, Landlord and Tenant Act reform, changes to the planning system and the affect on clients of the 2005 Rating Revaluation. It is also important to keep on top of economic and property market trends, as well as any case law relating the areas of work in which they are involved.

For a self-employed surveyor, time taken in keeping up to date with such developments in practice is at the expense of lost fee-earning time, while the broad level of overheads are still incurred. However, although full-day and half-day courses might consequently appear unattractive, they enable contact to be made with other surveyors, and also perhaps potential clients.

Self-employed surveyors do not benefit from the arrangements in larger firms where a training manager helps co-ordinate training needs, or a smaller firm where a graduate, for example, helps keep colleagues up to date with articles, and the distribution of course notes. Greater personal responsibility therefore has to be taken for being up to date with market changes. A lack of awareness could lead to mistakes in practice, including complaints from clients, and negligence claims.

The professional journals tend to cover the main issues, and *Estates Gazette*, *Property Week* and *RICS Business* should be read regularly. E-mail, as well as postal address registration with individual RICS faculties ensures that more detailed updates are available. This includes alerts to developments in practice which then enable the necessary information to be downloaded from RICS website or other sources. It may also be necessary to receive more specialist surveying/property publications, as well as trade journals from an industry in which the surveyor is involved on behalf of clients.

RICS CPD/lifelong learning rules allow qualified surveyors to undertake two-thirds of their CPD by way of private structured study. Although simply "reading *Estates Gazette*" would not count, examples of qualifying activities (and entries in CPD records) could include "Private study of 6 March and 13 March 2004 *Estates Gazette* Mainly for Students Landlord and Tenant update — 2 hours"; and "Private study of Advantage West Midlands CPD Papers on development appraisal and property transactions in the regeneration sector — 1 hour".

Course notes could be studied, and text books used, together with websites, and daily newspapers — really anything which facilitates "structured" learning, and therefore entries such as the above in surveyors' CPD records. The records could actually be requested by RICS as part of spot checks, although as indicated above, the minimum number of hours should be comfortably exceeded. Also, if surveyors are wishing to become fellows of RICS and benefit from the designation "FRICS", requirements include the commitment to CPD.

It is helpful if self-employed surveyors link up with each other for training and discussions. Responsibility for research could be shared out, with each surveyor providing a seminar and related study notes to other members of the group. Some regional RICS branches encourage such activity, such as the Small Business Forum run by RICS West Midlands CPD Foundation. Even if other regions do not have such initiatives, RICS branches could support their set up through notification in regional journals. It should also be possible to obtain details from RICS Business Services of other chartered surveyors acting as sole practitioners, or running small businesses who could be interested in being part of suitable study groups. In some cases, a surveyor becomes self-employed, and in working in the same locality in which previous employment was undertaken, has established contacts for networking and CPD. It may be possible to return to the company, or indeed link up with another company in order to use their library and other resource material.

VAT

This chapter initially explains the basics of VAT and the point at which it is necessary to register with Customs and Excise. It then explains the pros and cons of VAT registration, and concludes with a worked example of VAT accounting, as well as covering other issues.

VAT basics

VAT is payable on certain goods and services, and is collected by Customs and Excise.

Although as part of the VAT system businesses charge business customers/ clients VAT on their sales/turnover, and pay VAT similarly within items of expenditure, VAT is a tax raised primarily from consumers. This is because in simple terms, if a business is registered for VAT, and charges VAT on its sales/ turnover to customers/clients, it is able to recover the VAT incurred within its items of expenditure. In contrast, domestic consumers simply pay VAT on their purchases, and this cannot be recovered.

The VAT terminology is "outputs" for sales/turnover and "inputs" for expenditure. "Supplies" means either outputs or inputs, and relates to a transaction. "Taxable supplies" means supplies subject to VAT (standard rated or zero rated — see below), and "Taxable turnover" similarly means turnover subject to VAT. "Output tax" is the VAT on sales/outputs, and "input tax" is the VAT on expenses/ inputs.

Goods and services are either standard-rated, zero-rated, or exempt (although there is also a reduced rate — see below). VAT registration is necessary if annual turnover is above £58,000 (2004–05) — see Compulsory registration below.

Standard-rated

If goods and services are standard-rated, 17.5% is added to the sale price/fee/ outputs. As with most goods and services, surveyors' usual services are standard-rated. A VAT registered surveyor billing a client £2,000 fees, adds £350 in respect of VAT. The surveyor is able to recover the VAT incurred within its items of expenditure/inputs (such as stationery, telephone bill, and accountants' fees).

Zero-rated

If goods and services are zero-rated, VAT is 0%, and is not added to the price of the goods/services, but the business is still able to recover the VAT element within its items of expenditure/inputs. It is unlikely that a surveyor would provide zero-rated services (although an example could be the sale of literature, such as CPD packs which a surveyor has written). More common examples of zero-rated supplies are food and drink (from a shop — not a pub/restaurant), newspapers, children's clothes, and public transport. As indicated above, zero-rated supplies are included as taxable supplies/taxable turnover even though the rate is actually 0% (and therefore influences whether VAT registration is necessary).

Exempt

If goods and services are exempt, the sales/turnover/outputs are not subject to the addition of VAT, and it is not possible to recover the VAT element within items of expenditure/inputs. Exempt supplies do not therefore form part of taxable turnover. Again, it is unlikely that a surveyor would provide exempt services, although fees for lecturing at a university or delivering private tuition to students, are likely to be exempt (although it would be necessary to establish the precise rules in relation to the particular education/training activity). Rentals on property are exempt, although as surveyors are aware, a landlord could elect to waive exemption, and add VAT to the rentals (the benefit being to be able to recover the VAT element on expenditure/inputs which could not be reclaimed if not electing).

Reduced

There is also a "reduced" rate of VAT of 5% which covers supplies of fuel and power used in the home, and by charities. If surveyors work from home, as indicated on p 139 in respect of self-assessment, an allowance for heat and light would be, say, £300 for the year, and would not involve seeking to recover VAT on apportioned gas and electricity bills.

Partial exemption

A business which is "partially exempt" makes both taxable supplies (ie standard-rated and/or zero-rated) and also exempt supplies. This creates an issue of how VAT is recovered on expenditure/inputs, bearing in mind that there are likely to be items which cover both areas of work (such as stationery and telephone expenses covering all activities) and need to be directly allocated or apportioned.

However, as indicated above, a surveying business would typically provide only services which are standard-rated, and VAT accounting would be relatively straightforward — with all outputs being subject to VAT, and VAT being reclaimed on inputs (provided they are allowable/business expenses). There is also a *de minimis* rule which benefits a business making a small level of exempt supplies, and avoids allocation/apportioning.

Compulsory registration

It is not necessary for a self-employed surveyor/small business to register for VAT if the taxable turnover (standard-rated and zero-rated outputs) is less than £58,000 over a 12-month period (2004–05 — increased from £56,000 for 2003–04). If any capital assets such as property, vehicles or equipment have been sold, these do not count as taxable turnover.

The rules are that if at the end of any month, the total value of taxable supplies made in the past 12 months or less is more than £58,000 (within 2004–05), VAT registration is required. Registration is also required if the value of taxable supplies is expected to exceed £58,000 in the next 30 days alone (although unlikely to be relevant to a new surveying business).

The date of registration is the first day of the second month after taxable supplies for the past 12 months have increased above £58,000. If, for example, in June 2004, taxable supplies increased above £58,000 in the 12 months to 30 June 2004, the registration date would be 1 August. The business could, however, request an earlier registration date (which could be beneficial in order to recover VAT on expenses, but also involves adding VAT to outputs).

VAT registration details are available from *www.hmce.gov.uk* or the National Advice Service on 0845 010 9000, and if these details change, a local Customs and Excise office can be contacted for initial guidance. A straightforward Application for Registration form (Form VAT 1) asks for details such as the nature of the business, whether taxable supplies have already been made (ie whether the business is already running, and income is being earned) and the date required for VAT registration. Customs and Excise provides a certificate of registration, containing details, including the VAT registration number, and the period for accounting. Details of the local office are provided — noting that registration is to a registration office, and that a local office deals with the surveyor's VAT affairs thereafter. In the above example, Form "VAT 1" should have been sent to Customs and Excise by 31 July. If the business has just started, and taxable supplies have not yet been made, Customs and Excise may require proof that taxable supplies will be made, and suitable details (such as business proposals, copies of contracts or letters of engagement with clients, and copy invoices of acquisitions of equipment, etc). An accountant's letter should however suffice.

If the level of £58,000 has been reached, but it is considered that turnover over the next 12 months will not exceed £56,000, and sufficient explanation and proof can be provided to Customs and Excise, it is not necessary to register for VAT. Customs and Excise should still be informed, working to the above timescales, that £58,000 has been reached, but a Form VAT 1 is not completed. Advanced planning ensures that adequate time is available to obtain confirmation from Customs and Excise, bearing in mind that delays could cause complications with VAT administration, and accounting (especially if Customs and Excise are not satisfied with the explanation).

With a retail business, sales are usually regular, and turnover easy to monitor. With surveying services, invoicing is more irregular, with scope to delay invoicing. VAT accounting works on an accrued basis, with completed work being counted as turnover. Accountant's advice should be taken on the precise rules, and any issues arising in relation to the surveyor's range of services.

If there are delays in receiving registration details from Customs and Excise, there are arrangements whereby clients can still be invoiced without the VAT number. Guidance on this is available from Customs and Excise. Advanced planning should, however, avoid any such difficulties.

Further details in respect of initial VAT registration are provided in the Customs and Excise guide, *VAT Notice 700/1 Should I be registered for VAT?* It might be possible to recover the VAT on some inputs incurred prior to VAT registration, where the inputs are attributable to making taxable supplies. *HM Customs and Excise Notice 700 The VAT Guide — a guide to the main VAT rules and procedures* provides more detail.

Voluntary registration

Even if sales/turnover is below £58,000 (2004–05), it is possible to register for VAT. Some small businesses do this in order to give the impression that they are larger than is actually the case.

Voluntary registration could be backdated (including at the stage of compulsory registration), but it is unlikely to be beneficial for the small surveying business as income which had not been invoiced on the basis of VAT would be liable for VAT.

De-registering

A business which is registered for VAT could de-register if the taxable sales/ turnover is less than £56,000 (2004–05) over a 12-month period — such as a surveyor whose business has declined, or who is winding down to semi-retirement.

Informing clients of registration

It would be important for a surveyor becoming VAT registered to inform clients. As indicated above, clients who are themselves VAT registered are able to recover the VAT element and be unconcerned, but clients who are not VAT registered are effectively faced with a price increase. It could therefore be necessary to adjust the fee basis for some clients.

Other issues

VAT affairs become more complex if, for example, a business has been taken over as a going concern, and its taxable turnover influences whether it is necessary to register for VAT, or if trading takes place with companies overseas. An accountant's advice should be sought if any such factors are relevant.

Pros and cons of registration: issues for surveyors

The principal benefit of VAT registration to a business, as shown above, is that the addition of VAT to fees/outputs enables VAT to be recovered on expenses/inputs.

However, whether this is beneficial overall for a business depends, in particular, on the VAT status of customers/clients. This is because customers/clients who are not VAT registered, are not able to recover the VAT element on what, to them, is expenditure/inputs. They will regard VAT as an extra cost, which makes the business' services more expensive, and less competitive. If customers/clients are VAT registered, they will be unconcerned as to whether VAT is payable, as it will be reclaimed. Although there could still be a cash flow disadvantage to the client/customer, this is of relatively small effect, and unlikely to be of concern.

For a small local retailer, for example, all customers will be domestic consumers, and will not be VAT registered (except perhaps for occasional business purchases). Here, the retailer is likely to prefer to not register for VAT. If having to register for VAT, all turnover is subject to VAT (as opposed to turnover above £58,000). VAT registration would, however, enable VAT to be recovered on purchases/expenses/inputs.

For a self-employed surveyor advising small companies who are not registered, and/or advising private individuals, it is similarly disadvantageous to register for VAT. If clients are larger concerns who are VAT registered, it is advantageous to register for VAT (as they recover the VAT on the surveyor's fees, and the surveyor recovers VAT on the expenses/inputs of the business).

Another factor relevant to whether VAT registration is beneficial is the extent of expenditure incurred by a business, and the nature of the expenditure. Expenses which are subject to VAT could be relatively low, and the savings from VAT registration not particularly significant. This could well be the case with a small surveying business — and especially when the surveyor works from home with minimal set up and running costs. Employees' salaries are not, of course, subject to VAT.

If the VAT element on expenses/inputs is not recovered, it is the amount including VAT which can be set against income in determining profits and tax liability. For example, £1,000 plus VAT expenses for the VAT registered surveyor would see £1,000 costs set against income/tax, but for the non-VAT registered surveyor would see £1,175 costs set against income/tax, with a higher rate/40% tax payer, for example, effectively getting back 40% of the VAT element anyway (and a lower rate tax payer 22%).

The cost of accounting also has to be considered in determining whether to register for VAT voluntarily — although the surveyor should be able to handle much of the ongoing administration personally.

Connected interests

Surveyors should also be aware of connected interests rules. It would not, for example, be possible for a self-employed surveyor to trade under two different company names, having an annual turnover of £40,000 each, and escape VAT registration which would otherwise be necessary with a turnover of £80,000. Similarly, operating part of the business as a self-employed surveyor, and setting up a limited company, would not circumvent the need to register for VAT. Accountants' advice should be taken on the precise rules regarding connected interests, as different activities and different structures of ownership may enable interests to be subject to their own individual VAT arrangements.

If a sole trader/self-employed surveyor has a number of trading names, the VAT number is the same, but if operating as a sole trader while also running a limited company, and/or being involved in a partnership, each has its own VAT number. A different VAT number does not mean that the interests are unconnected.

Accounting periods

After an initial part period, VAT returns might, for example, be made in respect of the periods 1 February to 30 April, 1 May to 31 July, 1 August to 31 October and 1 November to 31 January. Customs and Excise send a form which needs completing and returning, together with payment for any amounts due, by the end of the month following the accounting period (ie 31 May, 31 August, 30 November and 28 February). This means that outputs and inputs falling in the period, as per the dates of invoices/receipts, have to be accounted for within each period — although see cash accounting below. (If necessary, Customs and Excise should be able to change the accounting period to align with the financial year of the business).

Cash accounting

The surveyor/business is able to administer VAT on the basis of "cash accounting" (provided taxable turnover is less than £660,000 (2004–05).

This means that only the outputs for which payment has been received during the period have to be accounted for — and likewise only inputs where payment has been made. This ensures, for example, that £5,000 plus £875 VAT, total £5,875 invoiced on 3 April, but not received by 30 April, does not have to be included in the 1 February to 30 April return, thus preventing cash flow difficulties in terms of paying Customs and Excise the VAT element which has not been received from the client. Cash accounting is similarly beneficial in the case of bad debts, as although invoiced, VAT not received is not included in the returns.

Also, a flat rate scheme is available, designed to simplify accounting, but which is not often taken up by businesses

Cash flow planning

The above factors mean that the timing of invoicing/outputs and receipt of payment can be managed to some extent in order to assist with cash flow — with the VAT element being held for up to four months if managed precisely (ie payment received on 2 August, is not remitted to Customs and Excise until 30 November). However, in delaying invoicing and receipt in order to hold the VAT element for longer, the actual fee element (which is more substantial) is also being delayed. Expenses/inputs could be paid for during the VAT period in order to recover the VAT element, but again, it could be preferable overall to delay payment because the amount which is net of VAT is more substantial.

VAT calculations

The calculation of VAT on outputs is a straightforward addition of 17.5% (such as £1,500 plus 17.5% VAT (£262.50) = £1,762.50. For expenses, the VAT amount should be stated on invoices/receipts, but the calculation to establish, for example, the VAT element and net amount in an expense of £850 would be £850 ÷ 1.175 = £723.40 — so VAT is £126.60 and the net amount £723.40 — ie £723.40 plus 17.5% VAT (£126.60) = £850.

Accounting for VAT

Accounting works on the basis of totalling up outputs and inputs, and their respective VAT elements. If VAT due on outputs is greater than VAT reclaimed in inputs, payment for the difference is made to Customs and Excise when returning the form. Set out below is an example which shows how straightforward VAT accounting could be managed. The initial information is extracted from the surveyor's running records of income and expenses, and the return is prepared for the period 1 November to 31 January.

Outputs	Fee	VAT	Total	Invoice no.	Invoiced	Received
Evans Ltd	£1,500	£262.50	£1,762.50	3–4/24	06/09/03	26/10/03
Marshalls	£2,000	£350	£2,350	3–4/25	19/09/03	04/11/03
Stone & Son	£1,100	£192.50	£1,292.50	3–4/26	21/11/03	22/12/03
Falcon Estates	£2,500	£437.50	£2,937.50	3–4/27	28/11/03	06/01/04
Richard Allan	£1,500	£262.50	£1,762.50	3–4/28	28/11/03	09/01/04
Sentence	£3,200	£560	£3,760	3–4/29	03/01/04	28/01/04
Davidson	£1,000	£175	£1,175	3–4/30	15/01/04	
Booker-James	£2,000	£350	£2,350	3–4/31	02/02/04	

Expenses	Gross	VAT	Net	Ref.	Invoiced	Paid
Staples	£375	£55.85	£319.15	17	21/10/03	24/10/03
Rogers	£587.50	£87.50	£500	18	12/11/03	03/12/03
Walls Milne	£23.50	£3.50	£20	19	19/12/03	19/12/03
TSW	£400	£59.57	£340.43	20	06/01/04	27/01/04
BT	£110	£16.38	£93.62	21	26/01/04	

On the basis of cash accounting, outputs/income, relevant to the period 1 November to 31 January would be Marshalls, Stone & Son, Falcon Estates, Richard Allan, and Sentence: total £10,300 plus VAT at £1,802.50. Evans' payment was received in the 1 August to 31 October period, and has already been accounted for in that period. Davidson and Booker-James is included in the 1 February to 30 April period (or later, when paid) as payments have not been received within the 1 November to 31 January period.

Expenses/inputs relevant to the period 1 November to 31 January would be Rogers, Walls Milne and TSW: total £860.43 plus VAT at £150.57 = £1,011. Staples relates to the previous 1 August to 31 October period, and BT will be included in the period in which the amount is paid. The above entries simply derived from the surveyor's own records. Although it would not be beneficial to list the suppliers' invoice numbers, the surveyor may reference expenses in a numbered way, hence the reference column above.

The above calculations come from the actual breakdowns from individual invoices/accounting entries, rather than revised calculations of gross amounts of income and expenses. This ensures that the actual VAT amounts collected and paid are included on the VAT return, and that overall, the business' accounts balance without the need for a balancing VAT entry. It is always nevertheless helpful to undertake such a cross-check calculation, as this would highlight any discrepancy in the above breakdowns.

VAT return form entries

The VAT return form sent to Customs and Excise will have the following entries:

VAT due in this period on sales and other outputs	(1)	£1,802.50
VAT due in this period on acquisitions from other EC Member States	(2)	None
Total VAT due (the sum of boxes 1 and 2)	(3)	£1,802.50
VAT reclaimed in this period on purchases and other inputs (including acquisitions from the EC)	(4)	£150.57
Net VAT to be paid to Customs or reclaimed by you (Difference between boxes 3 and 4)	(5)	£1,651.93

£1,651.93 is the payment required by Customs and Excise, together with the form, by 28 February.

"(3)" is the "total output tax" and "(4)" is the "total input tax". On the back of the form, further guidance is provided in respect of the above entries. Had the VAT being reclaimed been greater than the VAT due, the form would be completed and returned in the usual way, with payment from Customs and Excise for the difference being awaited.

Also on the return, information is needed about the total value of outputs and inputs (with the above figures being only VAT). For the above example, the form would be completed as follows.

Total value of sales and all other outputs excluding any VAT. Include your box 8 figure	(6)	£10,300
Total value of purchases and all other inputs excluding any VAT. Include your box 9 figure	(7)	£860
Total value of all supplies of goods and related services, excluding any VAT, to other EC Member States	(8)	None
Total value of all acquisitions of goods and related, Services, excluding any VAT, from other EC Member States	(8)	None

Client expenses

Where a surveyor incurs expenses/disbursements on behalf of a client, and which are recovered, the expenses are recovered at the gross of VAT price, with the client accounting for VAT as appropriate.

Surveyors' records

As part of the surveyor's records, a column of VAT period could be added to the initial table above in order to assist VAT administration. Although software packages are available, and accountants could alternatively handle VAT requirements, for the typical self-employed surveyor/small surveying business, VAT administration should be relatively straightforward, and could be handled personally with occasional guidance on particular points from an accountant, or Customs and Excise themselves. VAT on some items, such as entertainment and private motor vehicles, cannot be recovered.

The surveyor needs to retain records of the amounts which formed the entries on the VAT return, together with copies of invoices/receipts. There is an overlap with records retained for income tax and corporation tax, but as is shown above, details of income and expenses require specific treatment for VAT accounting, especially if working to the cash accounting basis. Also, VAT payments are due on invoicing (or payment if cash accounting) whereas for income and corporation tax, some income and expenses may not yet be invoiced, but still have to be included in profit and tax calculations.

Invoices

If a business is VAT registered, there are certain requirements in respect of information included on invoices (including the name and address of the business, its VAT registration number, the customer's/client's name and address, a description of the services supplied, and the breakdown of the fee and VAT element). This should be straightforward for surveying businesses having standard rated supplies, but would be more complicated for other types of business having different types of supplies.

Planning finances

The surveyor/business needs to plan finances to be able to meet the payment to Customs and Excise. Although under cash accounting, only the VAT amounts from outputs where payment had been received would be included in the payment, a business does necessarily, in practice, set the precise VAT amounts aside to accumulate for payment as part of the VAT return.

Delays, Customs and Excise investigations, and penalties

A new business has to monitor its trading performance in line with VAT registration requirements, and ensure that registration takes place at the correct point in time.

Customs and Excise may investigate a surveyor's/small business' accounts and general business affairs to ensure that the VAT is being administered correctly.

If surveyors/small businesses have allowed the need to register for VAT slip, they should take advice from an accountant on the best way to handle this. It is possible to backdate registration, and although the VAT system incorporates a system of penalties and interest charges, small errors should not be too costly, if at all punitive.

Delays in registering could mean that sales/outputs have been invoiced without the addition of VAT, when they should, in fact, be subject to VAT (and VAT due to Customs and Excise). If this is the case, it is necessary to re-invoice for the full amount, or if the original invoice has been paid, invoice for the VAT element. A surveyor with established clients who are VAT registered should not have problems, but a retailer, for example, will simply have taken many payments from customers and be unable to seek any further VAT amounts due. Where VAT is due on amounts where it was not added initially, and the VAT element cannot be obtained from a client, an adjustment is made, such as a £1,000 invoice which excluded VAT now becoming £851.06 plus 17.5% VAT of £148.94.

If it is discovered that errors have been made on previous returns, a subsequent return would make the necessary correction using boxes 1 and 4 in the section on VAT form entries above (provided the net amount is £2,000 or less). If the error is greater, Customs and Excise should be informed immediately.

Further accounting illustrations involving VAT are shown in Chapter 11, Accounting.

Accounting

11

When starting out as a sole trader, most surveyors work on a self-employed basis, receiving fee income and incurring expenses which at the end of the financial year determine the profits they have made — and therefore their liability for income tax and national insurance. Account also has to be taken of other income, such as interest on deposit accounts, dividends and property income. Partnership arrangements involve similar issues (including partnership accounts) and limited company status increases the extent of statutory compliance, including accounting information.

This chapter explains the basic principles of accounting in the context of sole trader and limited company status, and concentrates on the records maintained, and the statements prepared.

Use of accounting information

As well as meeting statutory requirements, good quality accounting records are beneficial if looking to sell the business, take on a partner, raise finance, or update lenders. The more detailed information included in a business plan, for instance, the better the impression conveyed, and the greater the chance of securing a sale, partner or finance on the best terms. Accounting information also enables the surveyor to regularly review the business costs being incurred, and highlight significant costs that could be reduced. Financial information similarly supports business development activity.

Up to date accounting information enables advantage to be taken of any tax planning opportunities — as opposed to simply adding up totals beyond the final year end, and finding that opportunities have been lost. Cash flow planning is also supported by accounting records, including for larger liabilities such as taxation payments due. If accounts are not prepared on time, and there are delays in making self-assessment returns, penalties and interest charges are imposed by the Inland Revenue. Good quality accounting records maintained by the surveyor minimise accountants' charges, and also benefit dealings with the Inland Revenue and Customs and Excise.

Accountants

Surveyors need an accountant. Once broadly aware of the accounting requirements, relevant tax rules and items which can and cannot be set against tax, surveyors can undertake much of the accounting personally, with accountants undertaking a general check at the year end, dealing with returns to the Inland Revenue, and providing any guidance and tax planning advice throughout the year. Similarly, while initial guidance is needed in respect of national insurance and VAT, its general administration is easily managed thereafter by the surveyor.

The tax rules are complex, and subject to change. As the business develops, and for example, a partnership or limited company operates alongside sole trader status, certain connected interests rules apply. Help with personal financial planning, including injury and illness cover as well as the right choice of investments, is also available from accountants.

As indicated in Chapter 5, when commencing as a self-employed surveyor, it is necessary to inform the Inland Revenue (even if a self-assessment return is already being filed). Class 2 national insurance payments are due immediately, and class 4 national insurance payments are made alongside self-assessment for income tax liability.

Record keeping

The Inland Revenue requires tax payers to maintain suitable records in order that their liability for income tax and national insurance contributions is accurately assessed, and similarly corporation tax in the case of limited companies. Likewise, Customs and Excise require records to be kept for VAT purposes. Records should be retained for the current year plus the previous six years.

Copies of all invoices are retained, and also invoices/receipts for expenses incurred. This includes capital items going back a number of years, and for which capital allowances are claimed (see p 139). If a receipt is not available for an item of expense, a suitable note should be made. However, expenses which are set against income need to be justifiable to the Inland Revenue. Reasonable amounts for heat and light (see p 139) and non-receipted items for occasional photocopying, for example, should not cause problems, but a high amount of non-receipted expenses, such as for newspapers and taxis are likely to cause difficulties.

If a receipt is not available to support income, this is not a particular problem, provided a suitable record is maintained (although it is preferable that invoices are prepared). Invoices are, however, required if surveyors become VAT registered.

Details need to be retained of previous employment, including salary, bonuses, benefits in kind, share scheme/share options arrangements, and redundancy payments. These could be part of the surveyor's first year in self-employment, and therefore included in the year end self-assessment return.

Cheque books, paying in books and bank statements need to be retained, together with interest certificates in respect of bank and building society deposits, and dividend vouchers from share/equity investments. Records should also be kept of any property income received, capital gains, etc.

The accounting information prepared for self-assessment is relatively straightforward. Whereas a limited company operates a designated business bank account, and produces a balance sheet as well as a profit and loss account (with accounts having to balance) sole traders sometimes manage with domestic bank accounts, and only need to evidence income, expenditure and profit as part of self-assessment returns.

Although it is, in theory, possible for the sole trader to simply bundle invoices/receipts, and add totals, it is preferable to produce an organised summary of income, expenses, capital expenditure/capital allowances, interest payments (for domestic as well as business accounts), dividends, etc. This is easily done on a basic Word™ file, or alternatively through spreadsheets or suitable software packages. As long as the accounting information is accurate, and the right information retained to meet Inland Revenue requirements, the method of summarising information is down to surveyors' preference.

Tax enquiries

The Inland Revenue does not raise queries against a self-assessment return when it is submitted (other than obvious omissions or errors in calculations, for example) and instead undertakes tax enquiries/investigations subsequently.

If surveyors receive a tax enquiry, accountants' advice should be sought. Records upon which the returns were based need to be assembled, together with anything else requested by the tax inspector. The more information provided to the Inland Revenue, generally the better — including explanations of anything which appears unusual, or where receipts, for example, have been lost. Reluctance to supply certain information creates a suspicion that something is untoward.

The Inland Revenue's tax inspectors are familiar with the norms for different types of businesses, and are able to identify any irregular items which could be part of tax evasion. Records of the sole trader's domestic bank accounts as well as business accounts are usually requested. In the case of expenditure such as payments to family members for work undertaken, it is be important that payments have actually been made, and are not simply accounting entries to save tax (and even if the money has been paid from the business account, it is not then paid back into a separate domestic account of the owner of the business). Queries could also be raised by the Inland Revenue with clients and suppliers.

The Inland Revenue either provides confirmation that all is in order, or alternatively requires the surveyor to attend an interview and deal with queries. Smaller discrepancies involve paying a revised liability for the year in question, plus interest, but if more serious discrepancies are uncovered, the Inland Revenue is likely to examine previous years' returns, and could present the tax payer with a substantial bill. Sufficiently serious misdemeanours result in prison sentences.

If additional tax is due, there is scope for penalties to be cushioned by the tax payer's co-operation with the enquiry. The amount of additional tax due is sometimes still negotiable. This would not be the case if a surveyor works from home and puts the purchase of a lawnmower through the business, but could be where there are points of interpretation (such as apportionment between domestic and business use). The Inland Revenue also supplies guidance on the rules relating to tax enquiries, and rights to appeal.

Accounting principles, and the sole trader's accounts

In order to show the basic accounting principles, an illustration is initially set out below of a sole trader/self-employed surveyor starting in business on 6 April 2003, and preparing the year's trading information up to 5 April 2004. These are the dates for the 2003–04 tax year, and the records of income and expenditure enable the self-assessment returns to be completed. The business is not VAT registered, nor are there any staff receiving a salary as the sole trader is able to undertake all administration personally. The surveyor works from a home-based office.

Summary of trading

The sole trader's summary of the year is as follows:

Cash/funding: A business current account was opened, and £15,000 deposited. During the year, the sole trader's drawings totalled £5,000.

Income: Clients were invoiced for £32,000, with £29,500 being received, and a further amount of £2,250 not having been invoiced in respect of completed work due for invoicing. The total income is therefore £34,250.

Expenses: Stationery was purchased at the outset for £575, and a subscription taken for *Estates Gazette* and *Property Week* totalling £300. Business cards and letterheads cost £300, and advertising £500. Internet and e-mail cost £200. An accountant was paid £350 for initial advice, and lawyers similarly paid £350. Professional indemnity insurance cost £600, and telephone charges amounted to £700. Business trips throughout the year involved hotel bills totalling £2,217. Lunches, drinks, etc, amounted to £950. Various information totalling £150 was ordered from RICS library and training/CPD events were attended, totalling £225. Maps and copyright, etc, totalled £200, and a further £500 was incurred on a range of printing jobs. Stamps were purchased on a number of occasions, and totalled £165. An RICS subscription of £390 was paid. A private car was used, and 13,657 business miles were recorded, with 700 miles at 35p per mile (£245) recovered from clients as disbursements. In addition to a variety of petty cash expenses totalling £150, partial records of non-receipted photocopying costs, and journals are available, with the total estimated at £250. In addition to the actual items of expenditure, £300 is to be claimed as an expense in respect of heat and light for home-working. Expenses total £14,041, including mileage — see later.

Capital items: Computer equipment, camera, etc, for the business was purchased for £3,000, and office furniture for £1,000.

Although the above amounts are conveniently stated to the pound, the company records would include entries to the nearest penny. Totals are rounded to the pound for inclusion in self-assessment returns.

Accounting and taxation issues arising

From the above information, the following accounting and taxation issues arise.

Capital allowances — A percentage of the cost of capital items, known as capital allowances, is set against income/profit. Whereas stationery, for example, is an expense set against income in a particular year, computer equipment or furniture are capital items set against income over a number of years. The computer and camera, at £3,000, is written down at 100%, and the furniture, at £1,000, is written down in the first year at 40% (and 25% per year thereafter). Capital allowances are therefore £3,400 (£3,000 and £400). £600 (ie the balance from £4,000) is carried forward to the 2004–05 tax year in order to similarly calculate capital allowances, with additional purchases of capital items being treated in the same way as above. The 100% is a benefit for small businesses only, and the 40% for small and medium sized businesses. Accountants will provide more details. Where surveyors have existing capital equipment, acquired in a personal capacity prior to becoming self-employed, they are able to introduce it to the business in a similar way to having to acquire the equipment. The effective purchase price must be market value, and capital allowances still need to be adjusted to take account of any joint business and domestic use.

Home expenses — Where a surveyor works from home, and wishes to set, say, £300–£500 against income as an expense for heat and light, domestic bills need to be retained. A suitable apportionment is needed between business use and domestic use of telephones, with all bills again being retained. All such amounts and apportionments need to be reasonable.

Disallowable Expenses — Drinks/meals in connection with client meetings, for example, cannot be set against income, although overnight accommodation and an evening meal and breakfast are allowable (provided it is genuinely in connection with business, as opposed to being weekend family breaks put through the books). As is shown below, expenses which are not deductible against income/tax, and are disallowable, are still included in the company's accounts — but adjusted accounts are required for the purpose of assessing tax liability. However, in practice, a self-employed surveyor does not necessarily bother retaining records of lunch, drinks, etc, knowing that they would not be tax deductible (and effectively taken out of their own pocket).

Vehicle expenses — In accordance with Inland Revenue rates, the mileage charge is calculated at 40p per mile for the first 10,000 miles, and 25p per mile for the remaining 3,657 miles: total £4,914.25. Invoicing has included disbursements (chargeable to clients) of £245 in respect of mileage at 35p per mile for 700 miles. £245 is deducted from £4,914.25 — total £4,669.25 (say £4,669). Had the 13,657 miles not included the 700 miles/£245 recovered from clients, 10p per mile/£70 would still have to be deducted from £4,914.25 as profit on expenses still counts as income. There are different ways of treating vehicle expenses and accountants will provide detailed advice if need be, but records of road tax, insurance, servicing, MOT, repairs, and fuel need to be retained, together with a log of journeys and mileage.

Work in progress — The sole trader's work in progress need not be included in the year's income. Accountant's advice should be sought on how to treat work in progress, noting that there could be flexibility as to which year to apportion income — and which could be the difference between higher or lower rate tax liability.

Inland Revenue self-assessment returns

The information needs to be accounted for correctly on the self-assessment returns. The Inland Revenue sends self-employed surveyors the form and accompanying guidance notes. There are various boxes to fill in, including the self-employment section. The forms take account of all types of businesses, and include boxes which are not applicable to self-employed surveyors providing professional consultancy services. The guidance notes are another source of information which helps self-employed surveyors increasingly understand the tax system, and be able to handle basic accounting matters on their own.

Income — £34,250 is entered against sales/business income (turnover) and is likely to also be the same figure for gross profit/loss. Other income/profits could include business bank interest, or a rent from sub-letting, but in the above example for the surveyor, this is not applicable.

Expenses — Expenses need to be listed under headings of employee costs, premises costs, repairs, general administrative expenses, motor expenses, travel and subsistence, advertising, promotion and entertainment, legal and professional costs, bad debts, interest, other finance charges, depreciation and loss/(profit) on sale and other expenses. The surveyor therefore categorises the £14,041 expenses into these areas, retaining records of how the amounts were apportioned. If there are grey areas regarding the apportionment of expenses among the categories, provided income is accurate, and expenses are items which are reasonably set against income, the profit figure is still the same, and the Inland Revenue is relatively unconcerned. It would not, however, be acceptable to hide certain items under incorrect categories in order to, hopefully, lessen the chance of questions being asked, or a tax enquiry pursued. Total expenses are calculated, and then net profit/(loss) — ie sales/business income (turnover) add other income/profits less total expenses. The net profit/(loss) here is £20,209 (£34,250 less £14,041).

Disallowable expenses — The lunches, drinks item of £950 is not an allowable expense, and should be listed in a further tax adjustments to net profit or loss section under disallowable expenses.

Capital allowances — From the capital allowances section on the form, capital allowances are added in the tax adjustments to net profit or loss (£3,400 as calculated above).

Net profit — The net business profit for tax purposes, upon which income tax and national insurance for the year is based, is £17,759. A summary is:

	£
Turnover	34,250
Expenses	14,041
	20,209
Add disallowable expenses	950
	21,159
Less capital allowances	3,400
Profit	17,759

This is, in effect, a basic profit and loss account.

Also on the self-assessment form, a section of income from UK savings and investments details bank and building society interest, and dividends (see pp 53 and 54 in respect of tax calculations). Tax free interest from ISAs/TESSAs does not need to be stated. There may also be property income. Other income is income unrelated to the surveyor's business, and is non-business income. Income tax liability on other income is at the rates applying to salary/business profit, but other income is not subject to national insurance payments.

Accounts for other purposes

The sole trader does not have to produce accounts for self-assessment other than those required on the returns to the Inland Revenue. However, the above accounts and profit figure are for taxation purposes, and this is not necessarily the same as the actual profitability of the business. Disallowable expenses, for example, are still a business cost, influencing the actual profitability of the business. As another example, capital equipment has a life longer than the one year assumed by the 100% write down, and could be written off by way of depreciation over a longer period in the company accounts.

If the surveyor produced full profit and loss accounts for separate business purposes, presentation and content could, for example, include income from different types of work, and a categorisation of all expenses. Other expenses could include property rent, rates, secretarial salary and employers' national insurance, and loan interest. The benefits of good quality, up to date accounting information, were summarised on p 135. An illustration of a balance sheet is shown on p 143.

Dates for tax returns

If the surveyor's tax return was filed before 30 September 2004, the Inland Revenue would calculate the amounts due. If submitted later, the Inland Revenue might still calculate the amounts due, enabling the surveyor to know the precise amount to be paid on 31 January 2005, but the Inland Revenue is not obliged to do this. If filing a return after 30 September 2004, the surveyor should, in theory, calculate the precise amount, and arrange payment, but in order to save undertaking the detailed calculations, and perhaps to keep an accountant's costs down, an approximate, slightly over-estimated, amount can be paid to the Inland Revenue. The Revenue in due course provides its calculations, and repays any amounts due to the surveyor/tax payer, together with any interest (although sometimes the Revenue holds smaller amounts on account).

Inland Revenue calculations

The calculations undertaken by the Inland Revenue to assess the income tax and national insurance due are based on the examples shown in Chapter 5. This included income tax bands for 2003–04 of 0% up to the personal allowance of £4,615, 10% for the next £1,960 and 22% for the next £28,540. 40% is payable above £35,115. For the sole trader in the above example, income tax liability on a profit of £17,759 would have been £2,656.48 — calculated as £4,615 at 0%, £1,960 at 10%, and the balance of £11,184 at 22%.

Class 4 national insurance contributions for 2003/04 are based on 8% of profits and gains between £4,615 and £30,940, and 1% of profits and gains above £30,940. Class 4 national insurance liability is £1,051.52, calculated as 8% of £13,144 (£17,759 less £4,615). Income tax and national insurance therefore total £3,708.

Had the surveyor's profits been £50,000, income tax liability would have been £12,428.80, calculated £4,615 at 0%, £1,960 at 10%, £28,540 at 22% and the balance of £14,885 at 40%. National insurance would have been £2,296, calculated as 8% of £26,325 (£30,940 less £4,615), plus 1% of £19,060 (£50,000 less £30,940). Interest and dividend calculations for the higher rate tax payer were shown on pp 53 and 54. The Inland Revenue's calculations show the breakdown, although presented slightly differently from the above. Tax payers should, however, check the calculations.

Class 2 contributions would also have been paid by the sole trader for the year at £104 (£2 per week/£26 quarterly payment, based on the 2003–04 figures).

Timing of tax payments, and cash flow planning

Surveyors need to establish when they should make payments to the Inland Revenue. In the first year of the above example, the first payment is not made until 31 January 2005 (which is also the last day that the tax return for the 2003–04 year can be made without incurring penalties). However, at 31 January 2005, part-payment also has to be made towards the 2004–05 tax liability. This is 50% of the tax liability from the 2003–04 year — although if a previous year's profits were relatively high, and the current year's profits look like being lower, the liability for such part-payments in advance of the year-end is reduced by submitting an appropriate form to the Inland Revenue, to include an estimate of tax payable. If underestimated, interest would be paid on any tax due. If profits are likely to be higher than in the previous year, tax returns are submitted in the usual way, and liability then calculated (ie there is not a need to separately inform the Inland Revenue).

In the above example, the surveyor's next payment would be due by 31 July 2005, and would represent the second instalment of the income tax and national insurance amounts due for 2004–05. A tax return may not, however, have been submitted for the 2004–05 year at this stage, and final settlement takes place only once returns have been made, and the precise liability for the year has been calculated. Settlement is then due by 31 January 2006. For the 2005–06 year, the surveyor would similarly make payments in January 2006 and July 2006.

In the example of the surveyor's first year profits of £17,759, creating a liability for income tax and national insurance of £3,708, payment would be made of £3,708 by 31 January 2005. Also due by 31 January 2005 would another £1,854 on account for the 2004–05 year. A surcharge of 5% is made by the Inland Revenue if payments are more than 28 days late.

For a new business and/or a business/surveyor growing in profitability, the above provisions could mean that relatively high tax liability is due, especially around January. It is imperative that accounts are prepared as soon as possible, and an accountant advises on approximate tax liability and timing.

Statements of amounts paid and amounts due are also provided by the Inland Revenue.

Tax planning

It was mentioned above that in keeping good quality records, it is possible to undertake tax planning prior to the end of the year. If for example, business profits and other sources of income are likely to exceed the 40%/higher rate tax band in 2004–05 as the business develops, it would be better to achieve more income, less expenses, less capital allowances and more profit in 2003–04 at 22% (even though there is a small cash flow disadvantage/loss of interest through paying a higher tax liability). The position with national insurance also needs to be considered, as this will be 1% in the higher rate band, and 8% in the lower rate band. Again, accountants' advice should be taken on the possibilities available.

Further comments in respect of tax planning were included in Chapter 5, Business Status and Tax issues.

Balance sheet

As an illustration of basic accounting principles, if the sole trader prepared a balance sheet for the end of the year, it would contain the inputs below. This includes the headings from the Inland Revenue's self-assessment form, although a balance sheet does not actually have to be provided. The balance sheet should, of course, balance, and acts as a cross-check on accounting (ie if an invoice has been missed in preparing accounts, but a payment has been made, the accounts will not balance). It also needs to be ensured that the expenses of home/heat and light, mileage and the rounded amount for petty cash/various have been accounted for accurately. This includes payments being made from the business account to the surveyor's domestic account regarding home/heat and light (as no actual expense is otherwise incurred by the business), and mileage (as the surveyor has funded the motor expenses from his domestic account, and again, no actual expense is otherwise incurred by the business). If payments have not been made by the year-end, they are classed as trade creditors/accruals (ie amounts owed). Fees not received are debtors, as are any prepayments made. The balance sheet in the Inland Revenue's self-assessment form covers all businesses, and parts are not applicable to professional services such as surveying.

Assets	£
Plant and machinery	600
Other fixed assets	–
Stock and work in progress	–
Debtors/prepayments/other	4,750
Bank/building society balances	26,428
Cash in hand	–
	31,778

Liabilities	£
Trade creditors/accruals	4,969
Loans and overdrawn bank accounts	–
Other liabilities	–
	4,969

Net business assets	**26,809**

Represented by (capital account)	£
Balance at start of period	15,000
Net profit/loss	16,809
Capital introduced	–
Drawings	(5,000)
	26,809

In understanding the concept of accounts, it is helpful to think how everything has to balance, and that there must be two elements to each transaction taking place.

Plant and machinery is net of capital allowances for the year (example of balancing: the reduction in assets is a reduction in profit, with the bank account being unaffected as expenditure is not actually incurred).

Debtors/prepayments/other is the fee income not yet received but included in the income for the year (example of balancing: income is added to profits despite not being received, so needs to be added to Debtors/prepayments/other).

Trade creditors/accruals is home/heat and light and mileage which are yet to be paid from the business account to the surveyor's account (example of balancing: expenses incurred are matched either by a deduction from the bank account or an addition to trade creditors/accruals).

The net business profit for tax purposes, upon which income tax and national insurance for the year is based, is £17,759, but for the balance sheet needs to be £16,809 as this represents the actual profitability of the business — with £950 being disallowable expenses for tax purposes. Income tax and national insurance is not an amount owing as a trade creditor/accrual, as this is the liability of the surveyor, not the business. If this was a limited company, corporation tax is a liability for the business, and included as trade creditors/accruals.

The bank/building society balance is checked by: bank balance at start (£15,000), add income received (£29,500), less drawings (£5,000), less capital expenditure (£4,000), less expenses incurred/amount actually paid — ie excluding home and mileage (£9,072) = £26,428.

Bank/building society balances correspond with the balance on the bank statement. If there was a cash balance, such as because an amount was withdrawn for petty cash, there are both bank/building society balances and cash in hand entries. The check on the cash in hand balance is the actual amount of cash spent.

VAT

As indicated in Chapter 10, the business has to register for VAT if turnover is above £56,000 for the 2003–04 year/£58,000 for the 2004–05 year. The business could opt to register voluntarily if the turnover was below these levels.

If the sole trader was VAT registered, income/turnover included in the accounts is unaffected, and is still £34,250. As allowable expenses (excluding RICS subscription, home/heat and light, and mileage) were subject to VAT, and capital allowances were subject to VAT, revised figures for the calculation of profit (£17,759 in the initial calculations) are needed. Allowable expenses inclusive of VAT total a gross £7,732, and for the company's own accounts and for self-assessment accounting are £6,580 — ie £6,580 plus VAT at 17.5% = £7,732 (the calculation is £7,732 ÷ 1.175). VAT on expenses of £1,152 is reclaimed from Customs and Excise throughout the year, in practice by deduction from the VAT on income which is due to Customs and Excise. VAT would actually be recoverable on the fuel cost, but this does not change the calculations here.

Capital expenditure is treated similarly, with £4,000 equating to £3,404 plus VAT at 17.5%. £596 is reclaimed from Customs and Excise, and £3,404 is used for the capital allowances calculations. Revised calculations are £2,553 at 100% and £851 at 40%, giving a total for capital allowances of £2,893.

VAT which has been reclaimed gives a gain of £1,748. However, this reduces the level of expenses, and similarly capital allowances, thus adding to the profit figure upon which income tax and national insurance is due. In the example, expenses are £1,152 lower, and capital allowances are £507 lower, meaning that profits are £1,659 higher. At 22% income tax, and 8% national insurance, total 30%, the additional amount due is £498. VAT registration has nevertheless created a net benefit of £1,161 in this year (ie ignoring capital allowances carried forward).

Limited company accounts

It was mentioned in earlier chapters that a limited company structure could be preferable, particularly for tax efficiencies. The following simple example assumes that the surveyor takes no money from the business (ie no salary or dividends) and is provided to show the nature of abbreviated accounts which would be submitted to Companies House for a company of this size (and which are publicly available).

As mentioned on page 137, whereas the self-employed surveyor working to the basis of self-assessment does not have to produce a balance sheet, nor generally balance the overall accounts, a limited company requires a balance sheet, with all items balancing at the end of the year. Also, more detailed accounts are required for the Inland Revenue than the abbreviated version needed for Companies House.

It is assumed that income/turnover is £76,000 and expenses are £20,000. The surveyor does not employ other staff. Income not yet received/debtors is £10,000 and expenses/creditors not yet paid is £3,000. Although income and expenses are included in the profit and loss account net of VAT as the business is VAT registered, there is likely to be an amount for VAT debtors/creditors in the balance sheet. This is because VAT will have been collected on income prior to the year end, but is not paid to Customs and Excise as part of VAT accounting until the end of the tax period, and when payments are due. Similarly, VAT may have been payable on expenses, but not reclaimed from Customs and Excise (in practice, not deducted from VAT received on income). It is assumed that VAT creditors is £1,500, comprising £2,000 VAT on income of £11,428 and £500 plus VAT within expenses of £3,357 (ie £2,857 + VAT).

Together with expenses/trade creditors of £3,000, total creditors is £4,500. However, a further element of creditors is the corporation tax due at the end of the year.

The actual profitability of the business is easily calculated at £56,000 in the simple example shown (£76,000 income less £20,000 expenses). Corporation tax is payable at 19%, equating to a tax liability of £10,640. Total creditors are therefore £15,140.

The financial year is assumed to be 1 April 2003 to 31 March 2004. From the authorised share capital, an amount would have been "issued" share capital, and £1,000 issued share capital is assumed, comprising £1,000 shares of £1 each. The surveyor is the only director, and a wife or husband the company secretary.

The abbreviated accounts would typically contain an initial headed page (such as Surveyor Ltd, Abbreviated Accounts for the Period 1 April 2003 to 31 March 2004), a contents page, and a summary of company information, including details of directors, the secretary, registered office, registered number of the company and the company's accountants.

An abbreviated balance sheet is included on the lines of the following:

Current assets:	£
Debtors	10,000
Cash at bank and in hand	51,500
	61,500

Creditors:	£
Amounts falling due within one year	15,140
Net current assets:	46,360
Total assets less current liabilities:	**46,360**

Capital and reserves:	£
Called up share capital	1,000
Profit and loss account	45,360
Shareholders' funds:	**46,360**

As a check on the accounts by examining the cash balance:

* The initial cash balance was £1,000.
* £66,000 income was received, net of VAT (£76,000 total income less £10,000 debtors).
* £11,550 VAT on income was received and then paid to Customs and Excise (£66,000 at 17.5% = £11,500)
* £2,000 VAT was received on income but not yet paid to Customs and Excise.
* £17,000 expenses were paid, net of VAT (£20,000 total less £3,000 unpaid/creditors).
* £2,975 VAT on expenses was paid but then reclaimed (£17,000 plus VAT).
* £500 VAT on expenses was paid but not yet reclaimed from Customs and Excise).
* The VAT element on £10,000 fees yet to be received, or on £3,000 expenses yet to be paid is not relevant, owing to cash accounting (see Chapter 10).

£1,000 plus £66,000, plus £2,000, less £17,000, less £500 = £51,500. This corresponds with the balance on the bank statement at the year end.

As with the self-assessment accounts for the self-employed surveyor, various balancing items can be examined in order to understand the accounting process. Examples include called up share capital also adding to cash at bank and in hand. Net current assets and total assets less current liabilities are the same, but if there were longer term liabilities, such as outstanding loans, this appears in the accounts between these two entries. A loan would add to liabilities as a debt, and add to assets as bank/cash balances (with loan repayments having the opposite effect). The profit and loss account entry of £46,360 is the retained profit, calculated as £56,000 profit, less £10,640 corporation tax at 19%. Had a salary been paid to the surveyor/director, this would have been included in expenses, and reduced profit before tax. Dividend payments would reduce the retained profit.

The abbreviated accounts include other information, such as comment on the company's entitlement to exemption from audit, comment on the company's members not having required an audit, the director acknowledging responsibility for keeping accounting records and for preparing financial statements which give a true and fair view of the company's affairs — all of which make reference to the

Companies Act 1985 and the relevant sections. It will also state that the abbreviated accounts have been prepared in accordance with the special provisions of Part VII of the Companies Act 1985 relating to small companies. Limited other explanatory details are provided, including authorised and issued share capital.

As mentioned above, full statutory accounts would be required for the Inland Revenue, and a company tax form completed (which is the equivalent to the self-employment tax return). The profit and loss account for the limited company uses specific headings, with notes to the accounts providing additional information.

While the sole trader should be able to deal with much of the accounting for self-assessment personally, accountants' advice is essential for limited companies and partnerships, and also in respect of overall tax planning.

Property Investment and Development Interests

As mentioned in earlier chapters, surveyors are naturally positioned to establish private investment and/or development interests alongside a surveying business. This could be part of longer term financial/pension planning, or the start of a small property business.

This chapter provides a basic overview of the issues to consider in respect of strategy, finance/gearing and risk.

Surveyors' circumstances

While it is helpful for a new surveying business to receive the full attention and focus of the surveyor during its early stages of development, it often takes time for business to be won, and dual surveying and investment/development interests work well for sole traders. Finances, however, could be too squeezed to embark on property investment as well as a surveying business. Nevertheless, as the surveying business establishes itself with sufficient profitability and cash flow, investment becomes more feasible.

If a surveyor has accumulated wealth, it is, of course, relatively easy to begin investing in property. Acquisitions are initially made without the need for finance/mortgaging, and finance/mortgaging is arranged if sufficient properties are acquired to warrant this. Sometimes however, self-employed surveyors with sufficient wealth prefer not to raise finance, and retain a low risk investment profile for the long term, even if this means acquiring fewer properties. The self-employed surveyor, especially in the early stages of a surveying business, usually finds it harder to raise finance for property investment, compared with the salaried surveyor enjoying a regular income.

Investment and development

For a surveyor establishing their own property business, it is more likely that investment is appropriate in the early stages, instead of development (such as the acquisition of a building plot to build a house, or a run down period-style office to convert to flats). For anybody commencing investment with constrained finances, and especially if requiring mortgaging, it is important to initially select properties which are relatively easy to let, and do not require expenditure on improvements. Rental income is then secured immediately, and covers interest repayments, as well as maintenance costs, management costs, service charges and any other expenses. Voids are, however, always a risk, and loan repayments also affect cash flow.

Residential and commercial

Although many small investors opt for residential properties, a general practice surveyor should be capable of also acquiring commercial investment interests. Residential properties, with their typically smaller lot size, are usually more appropriate in the early stages.

The intensity of management is reflected in the returns available for both residential and commercial investments — and the surveyor is well-placed to actively manage properties held locally, and achieve yields/returns in excess of general market norms. Professional residential lets in a "nice" area tend to see far lower yields, for example, than a house let on a room-by-room basis to students, or flats let in a poorer area where there is a relative strength of rental demand against demand/affordability for owner occupation. For commercial properties, the owner's/surveyor's particular property expertise should also help secure returns well in excess of market norms, particularly by way proactive asset management strategies.

Buy to let finance

Buy to let finance can be secured initially (at competitive rates) but if the number of properties increase considerably, a more commercial arrangement is likely to be necessary (and which typically involves higher rates of interest). Investors should look to secure a small spread of lenders, and not become beholden to an individual lender, which could lead to uncompetitive interest rates, the scope to recall loans, and profiteering. In using a number of lenders, different arrangements are more likely to be secured, such as a combination of fixed rate and variable rate loans, and interest-only loans, including different points in time at which capital has to be repaid. As illustrated in Chapter 3, it is also important to build good relationships with lenders, and a larger volume of business, together with close working, should help lower rates of finance and favourable other terms to be negotiated. Lenders are particularly keen to see that the venture is well-managed, with low risk.

Refurbishments and extensions

As the portfolio develops, and is securing a surplus of rental income over interest repayments and other costs, there is more scope to acquire properties requiring refurbishment/works (where as well as incurring interest repayments and works expenditure, rent is not received until the property is ready for occupation and tenants are found). Refurbishments of recently acquired properties are a way of adding significant value, as are refurbishments to existing tenanted properties — such as £10,000 expenditure increasing the rent by £1,500 and equating to a 15% yield, whereas market yields may be in the region of 7 to 8% if properties are acquired. This also, of course, increases capital value, facilitating refinancing (see below), with works expenditure, if repairs, being set against income/profits and reducing tax liability (see below). Significant value could also be realised from extensions, particularly where property prices are high — first, because the land is virtually free, and, second, because the cost of construction tends to be low in relation to rental value and added capital value.

Refinancing and rising values

As mentioned above, increased capital values provide scope for refinancing. However, another example of investment issues, and as has happened over recent years, is that an increase in house prices of, say, 50% over three years, while rents remain static, means that the 8% yield (£4,000 ÷ £50,000) is now 5.3% (£4,000 ÷ £75,000). Refinancing at the same interest rate means that higher interest repayments increase costs and reduce cash flow (while rents do not increase). Interest repayment costs would though be lower, for example, if the increase in house prices has been driven by lower interest rates, as has been experienced over recent years.

Even if finance costs are lower, the margin between yield and finance cost is squeezed, and even disappears. This makes further acquisitions unviable, at least in terms of income return, so while the capital gain appears welcome, the general rise in market values stifles the development of the venture. Some small/private investors in the residential market are, however, prepared to buy properties for their long term benefits, and are content for rents to simply cover the mortgage interest repayments.

Even if not refinancing, attention always needs to be given to the returns being made against current capital value, and not just the initial purchase price. Increased capital values decrease yields (assuming the rent remains the same), and it could be preferable to sell and use the profits more profitably elsewhere. If, of course, rents rise in the above situations, the position is more favourable.

Gearing, risk, and market downturn

As always, it is important to remain alert to investment risk, including risks relating to the outlook for the economy, the level of interest rates, housing market trends, buy to let market trends, voids, local factors. As surveyors are aware, in good market conditions, the raising of finance (and gearing — ie the proportion of debt to equity) generally helps higher returns to be made, but gives greater scope for losses when market conditions change for the worse, and/or when letting or other management difficulties are experienced.

If as a simple illustration, a surveyor has £50,000 cash, a single flat could be bought without mortgaging. Alternatively, five flats could be bought, totalling £250,000, using the £50,000 as equity. This is a loan to value ratio (LTV) of 80% (£50,000 equity, £200,000 debt, and £200,000 ÷ £250,000 = 80%). A 75% loan to value ratio allows only four flats to be bought (£50,000 equity, £150,000 debt, and £150,000 ÷ £200,000 = 75%). This is a good illustration of why some investors are anxious to maximise their borrowing (and why some lenders offer high LTVs in order to win business).

At a rental return (yield) of 8%, the investor buying the single £50,000 property secures an income of £4,000 in the first year, ie 8%. The investor buying five properties producing 8%, and paying 6% pa in interest, also commits the same £50,000 equity, but makes £20,000 rent (£250,000 × 8%) and pays £12,000 interest (£200,000 × 6%). This is an income return/profit of £8,000, equating to a return of 16% on the £50,000 invested. The ability to raise finance therefore doubles the income return from 8% to 16%.

If capital values increase by 20% in the year, the investor in the single £50,000 flat benefits by £10,000 and the investor with five properties benefits by £50,000 (£250,000 × 20%) and doubles their £50,000 equity. Furthermore, the increase in the value of the properties from £250,000 to £300,000, enables refinancing to take place. The additional £50,000 equity created, retaining an LTV of 80%, provides the investor with £40,000 which is available to acquire further properties, which again at an LTV of 80% is £200,000 worth of property. This is how in the right market conditions a small property investment business has the potential to grow rapidly — bearing in mind also that its principals/investors are able to add their own further capital as well as seeking finance from lenders. The investor also, of course, generally benefits over time from increases in rents.

As mentioned above, gearing has its risks. Cash flow, rather than lack of inherent profitability, is a potential downfall of any business. If in the above example, rental values weaken, say from the approximately £4,000 pa to £3,500 per property (approx. £80 per week to £70 per week) because of an influx of new investors and the increased supply of flats to let, and interest rates increase to 7%, total rent is £17,500, finance repayments are £14,000, and the resulting £3,500 difference accounts for a 7% income return on the £50,000 committed (as opposed to 16% in the above calculation). The tighter the margin between the rental yield and the cost of finance, the greater the sensitivity (and one reason why higher yielding/actively managed investments are more defensive against changes in rents or interest rates, among other factors). In the above example, house prices need to fall only by 20% from the £50,000 per flat/total £250,000 for the geared investor's initial outlay of £50,000 to be wiped out (at least on paper).

The examples show how in the right market conditions, substantial profitability is achievable — 16% rental return and 100% capital return on the equity invested. How investors and developers actually make their money from property does not tend to be highlighted in text books and the property press, and instead comment relates mainly to yields, and valuation. Although information is available via Investment Property Databank (IPD) and other outlets, including geared and ungeared returns, these tend to be broader market statistics relating to better quality property interests.

In the examples, costs, voids and other issues are still relevant, although the calculations are sufficient to make the principal points. The many fundamentals of residential and commercial property investment are obviously outside the scope of *Starting and Developing a Surveying Business*.

Taxation and accountancy issues

Rents from property acquired by a sole trader (or surveyor in salaried employment) are usually classed as investment income, and are included in the surveyor's self-assessment return to the Inland Revenue (see Chapter 11). Expenses are set against income in order to determine a net income/net profit upon which tax liability is based, and income tax liability is due in the current tax year. Investment/property income is not subject to national insurance contributions (unlike the self-employed surveyor's profits for the year, and indeed the salaried surveyor's pay for the year). If, however, the surveyor is trading rather than investing in property, this is the nature of the business, and profits are regarded as the usual business profits (and subject to national insurance as well as income tax). Accountants' advice should be taken on such points.

Capital gains tax liability also needs considering, with gains from trading being business profits subject to tax, and investment gains being subject to tax, but benefiting from an annual allowance (£7,900 for 2003–04 and £8,200 for 2004–05 — or effectively double if assets are held in the joint name of husband and wife). Accountants' advice should be sought on the scope for an owner's residence of a residential investment property to minimise or avoid capital gains tax on future sale, and on the potentially complicated calculations for capital gains tax.

Taxation is a loss of cash flow, and any tax which is avoided (not evaded), or deferred, is effectively a source of funding (at nil interest) upon which further returns could be made. Investment opportunities in the year include acquiring properties requiring maintenance and repair expenditure, or undertaking repairs at particular points in time, in order to minimise the year end taxation liability. Accountants' guidance should be sought on the distinction between works expenditure which is maintenance and repair, and set against income/tax, and capital expenditure, which cannot be — and is effectively a further element of capital acquisition. Capital improvements limit capital gains tax liability on any property disposals in due course, although this may be many years away, and not be particularly beneficial in the short term. There are, of course, grey areas between repair and improvement.

If property is held in a limited company structure, its profits are subject to corporation tax, with any additional liability for the principals/shareholders arising on the withdrawal of dividends (see Chapter 5). In the case of a higher rate/40% tax payer, limited company status helps profits to be retained in the business. For any surveyors considering beginning investing, it is important that taxation and accounting issues are worked through with an accountant at the outset.

As indicated in earlier chapters, good accounting records are helpful when seeking to raise finance. The strategy for a property investment business is also dependent on the likely view that lenders (and any investors) will take of the company/venture. It might be beneficial to delay acquisitions nearing the end of the financial year, and present a more favourable position regarding cash holdings/cash flow and borrowing in a balance sheet, thus showing greater prudence in the investment approach taken, and a lower overall risk profile. Cash flow is usually regular with property investment, but student lettings, for example, might see no rental income earned from the final term cheque in April, through to the return of students in September/October — with a concentration of voids also being more likely. Summer rents and advanced lettings, however, ease cash flow and risk. The financial year end for the purpose of company accounts could be set when cash flow is likely to be at its best.

Reflecting other interests

One aspect of private property investment is how people value their time. The buy to let market, for example, sees people acquiring residential property almost as a hobby and interest, and do not value their time as it is simply absorbed in leisure time. The self-employed surveyor needs to remain aware of the time taken personally in acquisition, letting, and management, as this is at the expense of lost fees (assuming that the business is busy enough). Such a factor becomes of greater significance as both the surveying business and the property business develops. Support from family and friends, and the instruction of agents, helps ease pressures and bring about optimum profitability overall.

The profits from a successful surveying practice may appear to dwarf the returns achievable from property investment. Investment is still likely to beneficial from a longer term perspective (because income from existing investments is earned over the years without involving much time commitment from the surveyor, whereas the surveyor usually always needs to work in order to generate business profits).

In respect of professional ethics, if surveyors are involved in investment and development activity in the same market in which they provide surveying services, various issues could emerge regarding conflicts of interest, and connected interests (see Chapter 9, RICS Requirements).

Further Information

Sources of further information are set out below, beginning with details of advisory support in respect of establishing a surveying business.

Advisory support

Marler Waterhouse, facilitated initially by Midlands Property Training Centre (MPTC), provides concessionary advisory support in respect of starting up a surveying business (although unfortunately not to employees while working for firms who are already clients of Marler Waterhouse).

Marler Waterhouse, which is run by Austen Imber, author of *Starting and Developing a Surveying Business*, specialises in business and training consultancy to the property sector, covering property consultancies, corporate and public sector property teams and investment and development companies. For prospective new surveying businesses, this includes viability assessments, business plans, marketing plans, business development support, accounting and financial issues, etc. For established businesses, Marler Waterhouse is involved in areas such as strategic review, business growth, and business management. Initial contact is to *startup@mptcentre.org*. Austen Imber can be contacted at *austen.imber@marlerwaterhouse.co.uk*.

Literature, and guidance notes

A range of text books, guidance notes and other literature can be found at the Estates Gazette Books website, *www.propertybooks.co.uk*. The Estates Gazette Interactive website, *www.egi.co.uk*, is a subscription service, which as well as daily news updates, contains an archive of articles and technical updates since 1986. RICS main website, *www.rics.org.uk*, provides news and various market updates, and economic bulletins.

CPD, lifelong learning

Courses, seminars, and study material are usually advertised through *Estates Gazette*, *Property Week*, and RICS. Advantage West Midlands, the regional development agency for the West Midlands, facilitates free of charge CPD papers which are particularly suitable for self-employed surveyors and smaller businesses, and include references to other free of charge material. Registration is to *smallbus@mptcentre.org*.

RICS compliance

Further information and queries in respect of compliance with RICS rules is available from RICS Professional Regulation and Consumer Protection Department. Tel. 020 7222 7000. Their section of RICS website is *www.rics.org/resources/standards*.

Inland Revenue

To register as being self-employed, the Inland Revenue's Helpline for the Newly Self-Employed is 08459 154515. Further information on setting up in business is available at *www.inlandrevenue.gov.uk/startingup*. A hard copy of the *Starting up in Business* brochure is also obtainable from 08459 154515. For businesses looking to take on a member of staff and having to administer PAYE (income tax and employees' national insurance deductions, and employers' contributions), the Inland Revenue New Employers Helpline is 0845 6070143.

Customs and Excise

To initially contact Customs and Excise, the National Advice Service is 0845 010 9000. The Customs and Excise website is *www.hmce.gov.uk*.

Companies House

The Companies House website is *www.companieshouse.co.uk* and the Companies House Contact Centre is 0870 3333636/*enquiries@companies-house.gov.uk*.

Business Link

Business Link is managed by the Department of Trade and Industry (DTI) and includes general guidance on a range of matters for small businesses. The website, *www.businesslink.gov.uk*, includes a section on employing people, which summarises employee rights and other employment issues.

Index

Cash
    accounting (VAT)... 130
    flow... 12, 29, 30, 42
    flow (income tax/NI) 142
    flow (VAT) 130, 131
Chargeable hours 5, 6
Class 1 national insurance 55
Class 2 national insurance 56
Class 4 national insurance 56
Clients... 11, 14, 65, 70, 87
Clients' accounts.. 107
Commercial finance/loans... 33
Companies House... 47
Company
    director.. 48
    formation.. 48
    names 47, 64, 112
    secretary.. 48
Complaints procedure 108
Compulsory registration (VAT)................................... 127
Conduct befitting 121
Conflicts of interest.. 120
Connected interests (VAT) 130
Continuing professional development 122
Corporation tax... 57
Costs (start up, general)......................... 5, 6, 29, 77, 80, 84, 92, 138
CPD 122
Crowded markets... 9
Customs and Excise...................................... 125, 127, 134

Data protection 115
Development (property).. 149
Director (company)... 48
Disallowable expenses (taxation) 139
Disciplinary action (RICS) 110
Dividends 54
Domestic mortgage.. 31

Economy 12
Employees' national insurance 55
Employers' national insurance 55
Entrepreneurs 17, 28
Examples of sole traders 5, 7, 9, 13, 18–21, 91–100, 138
Exempt (VAT) 126
Existing employment 15
Expansion 75–90